Bioinspired Superhydrophobic Surfaces

Bioinspired Superhydrophobic Surfaces

Advances and Applications with Metallic and Inorganic Materials

Frédéric Guittard
Thierry Darmanin

Published by

Pan Stanford Publishing Pte. Ltd.
Penthouse Level, Suntec Tower 3
8 Temasek Boulevard
Singapore 038988

Email: editorial@panstanford.com
Web: www.panstanford.com

British Library Cataloguing-in-Publication Data
A catalogue record for this book is available from the British Library.

Bioinspired Superhydrophobic Surfaces: Advances and Applications with Metallic and Inorganic Materials

Copyright © 2018 Pan Stanford Publishing Pte. Ltd.

All rights reserved. This book, or parts thereof, may not be reproduced in any form or by any means, electronic or mechanical, including photocopying, recording or any information storage and retrieval system now known or to be invented, without written permission from the publisher.

For photocopying of material in this volume, please pay a copying fee through the Copyright Clearance Center, Inc., 222 Rosewood Drive, Danvers, MA 01923, USA. In this case permission to photocopy is not required from the publisher.

ISBN 978-981-4774-05-5 (Hardcover)
ISBN 978-1-315-22961-4 (eBook)

 Printed and bound by CPI Group (UK) Ltd, Croydon, CR0 4Y

Contents

Preface vii

1. Introduction 1

2. Theoretical Part 9

3. Fabrication Processes 13

   3.1 Etching in Acidic Media 13
      3.1.1 Metals 13
      3.1.2 Electrochemical Machining 17
      3.1.3 Silicon 18
   3.2 Plasma Processes 21
      3.2.1 Plasma Etching 21
      3.2.2 Sputter Deposition 23
   3.3 Laser 27
   3.4 Etching in Basic Media 32
      3.4.1 Copper 32
      3.4.2 Others 37
   3.5 Anodization 38
      3.5.1 Aluminum 38
      3.5.2 Titanium 45
      3.5.3 Copper 47
      3.5.4 Others 48
   3.6 Electrodeposition 49
      3.6.1 Noble Metals 49
          3.6.1.1 Silver 49
          3.6.1.2 Gold 50
          3.6.1.3 Platinum and Palladium 51

		3.6.2	Non-Noble Metals	52
			3.6.2.1 Copper	52
			3.6.2.2 Others	56
		3.6.3	Rare Earths	61
		3.6.4	Sol-Gel Electrodeposition	62
	3.7	Electroless Deposition		62
		3.7.1	Silver	63
		3.7.2	Gold and Platinum	66
		3.7.3	Others	66
	3.8	Hydrothermal Processes		68
		3.8.1	Zinc Oxide	69
		3.8.2	Others	75
		3.8.3	Applications	76
			3.8.3.1 Photoluminescence	76
			3.8.3.2 Photocatalytic properties	77
			3.8.3.3 Others	78
	3.9	Use of Nanoparticles		82
		3.9.1	Nanoparticles	82
		3.9.2	Nanocomposites	86
		3.9.3	Colloidal Lithography	90
		3.9.4	Textured Substrates	92
		3.9.5	Aerogels	93
			3.9.5.1 Silica aerogels	93
			3.9.5.2 Graphene aerogels	94
	3.10	Chemical Vapor Deposition		97
		3.10.1 Carbon		97
		3.10.2 Silicon		102
		3.10.3 Carbides and Nitrides		103
		3.10.4 Oxides and Sulfides		105
4.	**Conclusion**			**109**
References				111
Index				183

Preface

Materials with superhydrophobic or related properties are one of the most studied subjects in the literature from a theoretical point of view and also for the large range of possible applications, for example, in anticorrosives, antibacterials, optical devices, or sensors. The study of natural species with special wettability showed the importance of surface structures and the surface energy of the resulting surface properties.

Various strategies can be used to reproduce superhydrophobic phenomena in laboratory. In this book, we focus especially on the use of metallic and inorganic materials. Indeed, they present unique properties, for example, in terms of thermal resistance, mechanical resistance, chemical and ageing resistance, optical (transparency, antireflection, photoluminescence), and electrical properties (conducting, semi-conducting, insulating). In this book, we review most of the strategies used in the literature using metallic and inorganic materials to obtain superhydrophobic or related properties.

Chapter 1

Introduction

Plants present on their cuticle various surface structures and chemical compounds responsible for their different wetting properties [1–7] (Figure 1). Among them, surfaces with superhydrophobic properties remain one of the most fascinating examples. Superhydrophobic surfaces are characterized by apparent contact angle (θ_w) and low water adhesion (low hysteresis H_w and low sliding angle α_w). The superhydrophobic and self-cleaning properties of the famous lotus leaves (*Nelumbo nucifera*) are due to hierarchical (double structured) surface structures composed of large convex cells and nanostructures hydrophobic wax crystals [2, 3] (Figure 1). Many other plants display superhydrophobic properties [4, 7]. A dual-scale or fractal surface roughness associated to intrinsically hydrophobic materials is one of the easiest ways to obtain superhydrophobic properties with high robustness, which is the stability of the superhydrophobic properties against high pressure. Robustness is very important, for example, for these plants to maintain their superhydrophobic properties during rainfalls.

Superhydrophobic properties are also present in many animals. Usually, fogging occurs when moisture condensation accumulates into droplets larger than 190 nm (1/2 of the shortest visible light wavelength). It can scatter light and highly reduce optical performance of surfaces. It was shown that the

---

Bioinspired Superhydrophobic Surfaces: Advances and Applications with Metallic and Inorganic Materials
Frédéric Guittard and Thierry Darmanin
Copyright © 2018 Pan Stanford Publishing Pte. Ltd.
ISBN 978-981-4774-05-5 (Hardcover), 978-1-315-22961-4 (eBook)
www.panstanford.com

eyes of mosquitoes and flies are not only superhydrophobic but also antifogging [8, 9], which provides clear vision in highly humid environments (Figure 2). These properties are due to highly ordered surface structures. The eyes are composed of hexagonal-close-packed ommatidia (≈20 µm long) and each ommatidium is covered by ≈100 nm bubble-like protuberances. The ordered arrangement of the eyes at both micro- and nanoscale is responsible for both superhydrophobic and antifogging properties. The loose arrangement of nanoprotuberances on hexagonal-close-packed ommatidia makes an extremely discrete and nonplanar triple-phase (liquid–air–solid) contact line. This contact line is energetically favorable for driving the spherical fog drops effortlessly from the surface.

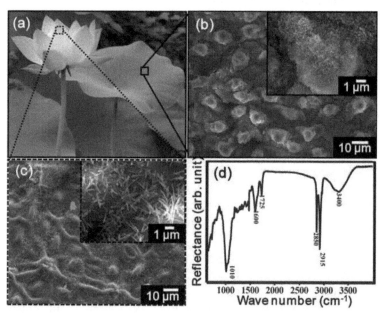

Figure 1 (a) Pictures of a lotus flower and a lotus leaf, SEM images taken from (b) the lotus leaf and (c) the flower surface; (d) ATR-FTIR spectrum recorded from the top of a lotus leaf surface. Reprinted with permission from ref. 4, copyright 2011, Royal Society of Chemistry.

Many insects such as water striders are also able to slide on the water surface almost without any effort [10–14] (Figure 3). Indeed, while the world of humans is governed by gravity, that of water-walking arthropods is dominated by surface tension.

In order to combine buoyancy and ultra-low adhesion to water surface, the feet of these insects are covered by highly dense hairs (30 μm in length and about 1 μm in base diameter) composed of hydrophobic waxes. Moreover, some of these insects are able to resist submersion underwater and are also able to breathe underwater by forming a plastron (air trapped by their hairs) against hydrostatic pressures.

Figure 2 Pictures of an antifogging and antireflection fly eye at different scales. Reprinted with permission from ref. 8, copyright 2014, Wiley.

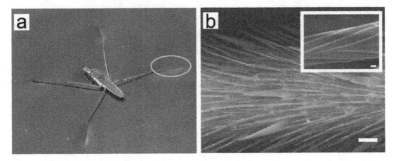

Figure 3 Picture of the leg of a water strider capable of walking on the water surface. Reprinted with permission from ref. 12, copyright 2010, American Chemical Society.

Jiang et al. were the first to observe that rose petals possess both extremely high water contact angles above 150° but also high adhesion in opposition to the classical superhydrophobic properties [15] (Figure 4). The term "petal effect" is often used in the literature [15-17] to describe these properties while Marmur proposed the term parahydrophobic [18]. They observed that these properties are due to the presence of ordered micropapillae (16 μm in diameter and 7 μm in height.) with the presence of nanofolds on their surface. Here, water can enter the large spaces between the large micropapillae, but not inside the nanofolds. The same research group also observed that xerophate peanut *Arachis hypogaea*, which survives in arid and semi-arid regions characterized by high temperature and low rainfall, has leaves with extremely strong water adhesion [19] (Figure 4). Here, the authors observed that the leaves are covered by nanoslices with about 40–60 nm in thickness and microscale length. Hence, the presence of only nanostructures can lead to high water adhesion [20]. The gecko feet have also high adhesive forces toward water [21]. The feet are covered by high-density nanopillars giving a $\theta_W > 150°$ and an adhesive force of 66 μN. Thanks to these properties, the gecko is able to move on vertical substrates. The peach skin was also studied. It was found that the skin is covered by long and short hairs. The long hairs are mainly composed of hydrophobic waxes and the short hairs of hydrophilic polysaccharides, showing the importance also of the surface energy. The authors reported a $\theta_W = 142.3°$ and an adhesive force of 87.3 μN.

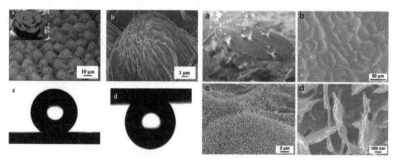

Figure 4 On the left, SEM images of red roses and on the right SEM images of peanut leaves with strong water adhesion. Panel on the left reprinted with permission from ref. 15, copyright 2008, American Chemical Society. Panel on the right reprinted with permission from ref. 19, copyright 2014, Wiley.

Watson et al. also studied the wettability of 15 species of cicadas and observed that the water contact angle is ranging from 76.8° to 146° with extremely strong water adhesion [22, 23] (Figure 5). These properties are due the presence of ordered nanodomes with different diameters, spacing, and height.

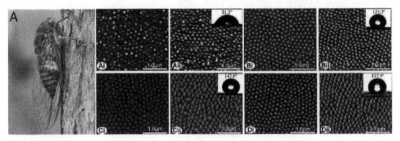

Figure 5 SEM images of different species of cicadas [22, 23].

Other insects such the embiopteran *Antipaluria urichi* are able to weave silk nanofibers with extremely high water adhesion [24] (Figure 6). These properties are extremely interesting in order to trap water droplets on the fiber surface. Here, Yarger et al. showed that these properties are due to the presence of loosely woven nanofibers.

Figure 6 Picture of the embiopteran *Antipaluria urichi* producing silk with strong water adhesion. Reprinted with permission from ref. 24, copyright 2016, American Chemical Society.

Water being vital, many animals and plants have developed strategies to capture water even in arid environments [25–30]. Among them, surfaces with contrasts in surface energy or roughness are largely employed. Shirtcliffe et al. showed that plants such as *Alchemilla mollis*, *Echeveria*, *Lupin regalis* and *Euphorbia* have parahydrophobic leaves with sticky zones to collect water droplets but with highly hydrophilic central zones to guide the water droplets when their size becomes critical [25] (Figure 7). The authors also demonstrated the possibility of guiding water by creating superhydrophilic grooves surrounded by superhydrophobic walls. The Namib Desert beetles *Stenocara* sp. use also a similar strategy [28–30].

Figure 7 Pictures of different plants capable of guiding water droplets. Reprinted with permission from ref. 25, copyright 2009, American Chemical Society.

Other species use anisotropic roughness to guide to water [31–38]. This is the case of the butterfly *Morpho aega* [31]. The surface of its wings is composed of microscales overlapping in only one direction. Hence, a water droplet placed on the wing can roll off the surface if it is inclined in the direction of the scales but is pinned in the opposite direction. Moreover, the presence of microscales is also responsible for the blue color of this butterfly. The group of Jiang also reported that the cactus *Opuntia microdasys* from the Chihuahua Desert can also collect water from fog thanks to its spines [34] (Figure 8). Its spines contain microgrooves with a higher roughness near the tip than near the base leading to a wettability gradient. Moreover, inspired by cactus spines, the authors also fabricate fog collection systems originating from the combination of a Laplace pressure gradient and the wettability difference [35, 36]. The endemic Namib desert grass *Stipagrostis sabulicola* is also able to irrigate itself with fog water [37].

Introduction | 7

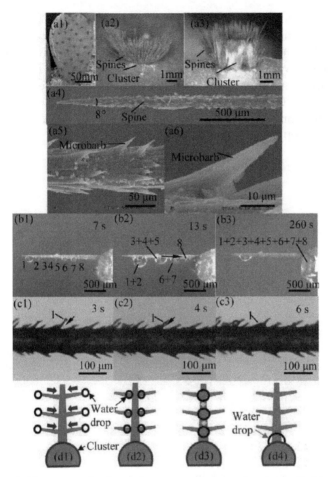

Figure 8 Pictures and schematic representation of water collection by cactus spines. Reprinted with permission from ref. 36, copyright 2014, American Chemical Society.

Usually low-surface-tension liquids such as oils have an extreme tendency to spread on any substrates. However, wingless arthropods of springtail class (*Collembola*) can repel many kinds of liquids even at elevated pressure, can resist bacterial adhesion, and also have high mechanical robustness [39–45] (Figure 9). These exceptional properties are especially due to the presence on their cuticle of extremely ordered surface textures with re-entrant curvatures such as mushroom and serif T structures. Moreover, the surface chemistry is also an important parameter.

It was shown that their cuticle is composed of different substances such as chitin, lipids, steroids, terpenes, and proteins. For example, the primary granules and interconnecting ridges, together forming the epicuticle or cuticulin layer, are mainly composed of proteins with a high amount of glycine, tyrosine, and serine. The composition of the amino acids is close to known structural proteins such as fibroin, collagen or resilin that often combine stiffness and toughness.

Figure 9 Picture and SEM images of the cuticle of superoleophobic springtails with reentrant structures. Reprinted with permission from ref. 39, copyright 2016, Royal Society of Chemistry, and from ref. 40, copyright 2014, The Royal Society.

Hence, it is first very important to understand the theories of superhydrophobicity before reproducing the surfaces. Hence, the next part is dedicated to the theoretical part.

Chapter 2

Theoretical Part

To describe the superhydrophobic phenomena, it is first fundamental to well understand the Young equation, which gives the apparent contact angle (θ^Y) of a liquid droplet on a "smooth" substrate (Figure 10). This contact angle depends on three surface tensions: the solid–vapor (γ_{SV}), the solid–liquid (γ_{SL}), and the liquid–vapor (γ_{SV}) interface.

Figure 10 Schematic representation of a water droplet on a smooth surface following the Young equation.

The presence of surface structures or roughness is necessary to obtain superhydrophobic properties [46, 47]. Indeed, two equations are very often used to explain the effect of surface roughness on the surface wettability. When a water droplet

---

Bioinspired Superhydrophobic Surfaces: Advances and Applications with Metallic and Inorganic Materials
Frédéric Guittard and Thierry Darmanin
Copyright © 2018 Pan Stanford Publishing Pte. Ltd.
ISBN 978-981-4774-05-5 (Hardcover), 978-1-315-22961-4 (eBook)
www.panstanford.com

follows the Wenzel equation $\cos \theta = r\cos\theta^Y$, where r is a roughness parameter, the liquid enters all the surface roughness in contact leading to a full solid–liquid interface [46] (Figure 11a). Following this equation, it is possible to obtain extremely high θ but only if $\theta^Y > 90°$. This is the case, for example, of some intrinsically hydrophobic polymers such as polytetrafluoroethylene (PTFE) or polydimethylsiloxane (PDMS). Moreover, with this equation the water adhesion or hysteresis is usually important because the roughness parameter increases the solid–liquid interface.

However, it was also reported in the literature that superhydrophobic properties can be reached from intrinsically hydrophilic polymers [48]. This possibility is not allowed with the Wenzel equation but can be predicted with the Cassie–Baxter equation [47]. In this equation, air is trapped between the droplet and the substrate inside the surface roughness (Figure 11b). The Cassie–Baxter equation is $\cos \theta = r_f f \cos\theta^Y + f - 1$, where r_f is the roughness ratio of the substrate wetted by the liquid, f the solid fraction and $(1-f)$ the air fraction. With this equation, it is possible to have superhydrophobic properties with very low adhesion, especially if the air fraction is very important.

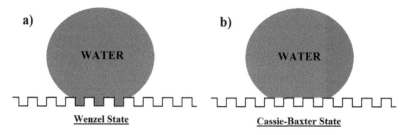

Figure 11 Schematic representation of a water droplet on a rough surface following (a) the Wenzel equation and (b) the Cassie–Baxter equation.

Several authors demonstrated that the presence of two-level hierarchical surface (micro and nano, for example) is very important to maintain stable or robust superhydrophobic properties by enlarging the energy difference between the Cassie–Baxter and the Wenzel states [49–52]. For example, Huang et al. showed that the first-level structure (nano) allows the surface to sustain the highest pressure found in nature while the second-level structure (micro) leads to a dramatic reduction in the contact area and hence minimizes adhesion between water and the solid surface [49].

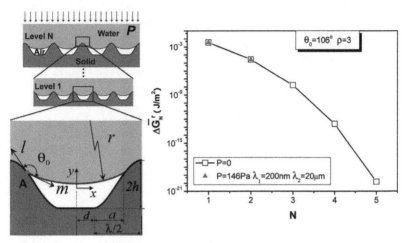

Figure 12 (Left) A liquid drop in contact with an N-level hierarchical wavy surface. (Right) Relationship between the reverse energy barrier and the structure level N. Reprinted with permission from ref. 49, copyright 2010, American Chemical Society.

The Cassie–Baxter equation can also lead to both high θ_w and high water adhesion if the air fraction is less important, as observed on rose petals, for example. This can be possible by creating surface structures able to trap a lower amount of air (for example, horizontally aligned nanofibers) and/or by increasing the surface energy (γ_{SV}). Indeed, Bhushan and Nosonovsky [16] demonstrated that many intermediate states between the Wenzel and the Cassie–Baxter state can exist with high θ_w and high water adhesion as shown in Figure 13.

The Cassie–Baxter equation can also be used to predict superoleophobic properties even if the materials are intrinsically oleophilic ($\theta_{oil}^Y < 90°$) [53–57] (Figure 14). The main difference is that the oils have a much lower surface tension (γ_{LV}) and as a consequence a much higher tendency to spread on any substrate. Moreover, it was demonstrated that one of the ways to produce superoleophobic surfaces is to fabricate surface structures with re-entrant curvatures as represented in Figure 14 [54, 55]. Indeed, the presence of re-entrant structures allows a high pinning of the three-phase contact line of oil droplets and as a consequence can highly increase the energy barrier between the Cassie–Baxter and the Wenzel states.

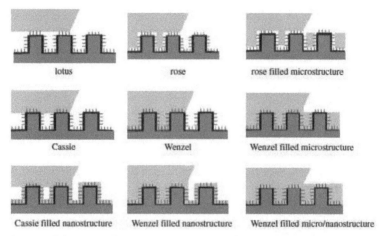

Figure 13 Schematic of nine wetting scenarios for a surface with hierarchical roughness. Reprinted with permission from ref. 16, copyright 2010, The Royal Society.

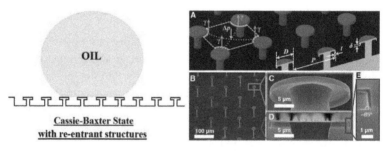

Figure 14 On the left, schematic representation of an oil droplet on a rough surface with re-entrant structures following the Cassie–Baxter equation and on the right SEM images of superoleophobic fabricated micro-pillar substrates with doubly re-entrant structures. Reprinted with permission from ref. 54, copyright 2014, the American Association for the Advancement of Science.

After the description of the superhydrophobic phenomena, the next section will be dedicated to the main fabrication processes to obtain superhydrophobic surface using metallic and inorganic materials.

Chapter 3

Fabrication Processes

3.1 Etching in Acidic Media

3.1.1 Metals

When a non-noble metal substrate is put in contact with acids (H$^+$), it can react following an oxido-reduction reaction: M + H$^+$ → M$^+$ + (1/2) H$_2$.

A metal substrate can be etched if its oxidation potential ($E^{Ox}_{M^+/M}$) is below than $E^{Ox}_{H^+/H_2}$. Moreover, etching rapidity is dependent, for example, on the difference $\Delta E = E^{Ox}_{H^+/H_2} - E^{Ox}_{Mn^+/M}$ and the H$^+$ concentration. If an oxide layer is present on the metal substrate, H$^+$ can also dissolve this oxide layer before to attack the metal. Different oxidation states for the metal often exist.

It is known that the etching of aluminum substrates by HCl leads to microporous "building block" architectures [58–65] (Figure 15a). After treatment with fluorinated materials, it was possible to obtain directly superhydrophobic properties (θ_w = 161.2°, α_w < 8°).

It is possible to control the surface properties with HCl concentration and etching time, for example. Moreover, the substrates displayed also anticorrosion and anti-icing properties. In order to obtain micro- and nanostructured superhydrophobic properties with very low adhesion and also superoleophobic properties, these etched substrates can be modified to induce nanoroughness on the "building block" microstructures. Different

---

Bioinspired Superhydrophobic Surfaces: Advances and Applications with Metallic and Inorganic Materials
Frédéric Guittard and Thierry Darmanin
Copyright © 2018 Pan Stanford Publishing Pte. Ltd.
ISBN 978-981-4774-05-5 (Hardcover), 978-1-315-22961-4 (eBook)
www.panstanford.com

strategies were used in the literature. Bhushan et al. reported a nanoetching process in the presence of HNO_3 and $Cu(NO_3)_2$ in order to form nanopores on the "building block" microstructures [61]. The substrates displayed superoleophobic properties with $\theta_{hexadecane} = 152°$, $H_{hexadecane} = 12$ and $\alpha_{hexadecane} = 14°$, and also self-cleaning properties and mechanical durability. Zhou et al. treated them with boiling water in order to create petal-like crystalline boehmite ($\alpha \cdot Al_2O_3 \cdot H_2O$) nanostructures with superhydrophobic properties and very low adhesion ($\theta_w = 161°$, $\alpha_w = 3°$) [64]. Huang and coworkers reported the formation of MnO_2 coral-like nanostructures by treating these aluminum substrates with $KMnO_4$ (Figure 15b), resulting in both superhydrophobic and anticorrosion properties [62, 63]. By contrast, if a carboxylic acid ($CH_3(CH_2)_nCOOH$) is added into the solution, the released Al^{3+} can react with $CH_3(CH_2)nCOO^-$ ions to form $Al(CH_3(CH_2)_nCOO)_3$ nanoclusters. Others researchers used nanoparticles to induce nanoroughness [66–68]. Finally, superoleophobic substrates were also reported by anodization process in oxalic acid on the microstructured aluminum substrates resulting in the formation of aluminum oxide nanowires. The substrates could repel low-surface-tension liquids as low as 21.1 mN/m (octane).

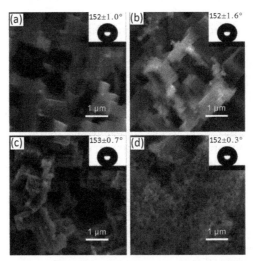

Figure 15 Aluminum substrate after etching in HCl (a) and post-treatment in $KMnO_4$. Reprinted with permission from ref. 63, copyright 2015, Royal Society of Chemistry.

The chemical etching of other metal substrates was studied in the literature [69–76]. In the case of copper substrates, the group of Attinger used different strategies [69]. First, copper substrates were etched in the presence of HCl and H_2O_2 in order to facilitate the etching: $Cu + 2HCl + H_2O_2 \rightarrow CuCl_2 + 2H_2O$. Copper substrates were also etched in the presence of $FeCl_3$: $Cu + FeCl_3 \rightarrow CuCl + FeCl_2$. These two strategies led to micropillars with extremely high water adhesion ($H_w \approx 150°$). Brass substrates were also etched with HCl and $FeCl_3$ leading to rough structures [71] (Figure 16). After heat treatment at 350°C, micro-nanolamellas of Cu_2O and ZnO were observed while using a final treatment with stearic acid, superhydrophobic flower-like structures of copper and zinc stearate complexes were obtained.

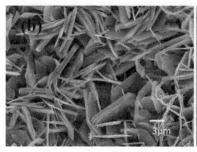

Figure 16 Brass substrate after immersion in HCl and $FeCl_3$. Reprinted with permission from ref. 71, copyright 2016, Elsevier.

The immersion of copper substrates in ethanol solutions containing a carboxylic acid such as *n*-tetradecanoic acid led to the formation of $Cu(CH_3(CH_2)_{12}COO)_2$ clusters consisting of flower-like structures composed of nanosheets [72–76]. The dimension of the nanosheets could be controlled by the concentration of *n*-tetradecanoic acid. After 3 days of immersion time, the surfaces displayed superhydrophobic properties ($\theta_w = 163°$). Zhang et al. showed the possibility of reaching nanoslices of hydroxy cupric phosphate heptahydrate $(Cu_8(PO_3OH)_2(PO_4)_4 \cdot 7H_2O)$ single crystals by immersion of copper substrates in H_3PO_4 and H_2O_2 solutions [76]. Their self-assembly induced the formation of porous microspheres while their formation and dimensions were highly depending on the phosphate ions, the immersion duration, and the growth rate on the H_2O_2 concentration. After post-treatment, superhydrophobic surfaces with very low hysteresis and sliding angle (θ_w = 169.6°, H_w = 4.2°, α_w < 5°) were obtained due to the combination of microspheres and nanoporosities.

For magnesium, different etching agents were reported in the literature [77–81]. For example, the etching of Mg substrates in $CuCl_2$ or H_2SO_4/H_2O_2 led after immersion in stearic acid to petal-like nanostructures with superhydrophobic and anticorrosion properties [77, 78]. Flower-like structures were also observed after immersing Mg substrates in alkylphosphonic acids or carboxylic acids [79, 80]. The surface morphology and hydrophobic properties could be controlled with the immersion time and also with the alkyl chain length. Xu et al. treated zinc substrates in HCl solution resulting in rough microstructured substrates [82]. After immersion in hot water, ZnO nanorods were formed leading after post-treatment with a perfluorinated silane to superhydrophobic and oleophobic properties. Zhou et al. also obtained superoleophobic properties of aluminum substrates after etching in HCl and immersion in boiling water [83]. They observed that the re-entrant structures obtained after etching in HCl were necessary to obtain superoleophobic properties ($H_{hexadecane}$ = 8.0° and $\alpha_{hexadecane}$ = 7.2° or H_{decane} = 45.1°, and α_{decane} = 40.1°) while the formation of petal-like nanostructures after immersion in boiling allowed them to decrease the contact angle hysteresis. By immersion of zinc substrates in the $CF_3(CF_2)_8COOH$ solution, $Zn(CF_3(CF_2)_8COO)_2$ nanosheets were obtained with superhydrophobic and superoleophobic properties with θ_w = 158.1° and $\theta_{rapeseed\ oil}$ = 155.6° [84].

Another strategy used in the literature involves the production of acids in situ, which could be obtained by the replacement of acids with substituted-trichlorosilane. Hence, the hydrolysis of substituted-trichlorosilane leads to silanols functions and released HCl in the solution. The group of Boukherroub showed that the insertion of zinc substrates in hydrolyzed perfluorotrichlorosilane induces the formation of zinc oxide (ZnO), simonkolleite [$Zn_5(OH)_8Cl_2.H_2O$], and zinc hydroxide [$Zn(OH)_2$] structures [85, 86]. The surface morphology was highly dependent on the hydrolysis rate of perfluorotrichlorosilane. They produced spherical particles of about 1.5–2.0 mm by controlling this parameter and obtained superhydrophobic surfaces with θ_w = 151°.

Finally, the etching of stainless steel substrates in HCl led to rough substrates [87]. Otherwise, multiscale roughness with nanoflake structures was reported by etching with a mixture of HF and H_2O_2, HNO_3 or $CuCl_2$ [88–91]. After post-treatment, superhydrophobic properties were obtained. Bormashenko et al. fabricated superoleophobic hierarchical stainless steel meshes by etching in HCl and modifying them with a fluorinated silane [92]. Liu et al. also etched stainless steel meshes but in the presence of HCl, $CuCl_2$, and stearic acid [93] (Figure 17). Dendritic structures of copper stearate were observed on the surface of the meshes. The resulting meshes displayed superhydrophobic and superoleophilic properties and could be used to separate oil–water mixtures.

Figure 17 Stainless steel meshes substrates after immersion in HCl, $CuCl_2$ and stearic acid and their use for oil/water separation. Reprinted with permission from ref. 93, copyright 2016, Elsevier.

3.1.2 Electrochemical Machining

Electrochemical machining was also used in the literature [94–98]. In this process, a metal substrate is electrochemically etched to created rough substrates in the presence of a highly corrosive electrolyte. For example, it was shown that superhydrophobic aluminum substrates with microporous "building block" architectures could be prepared in NaClO$_3$ electrolyte and at constant current density [94] (Figure 18). Moreover, the surface morphology and hydrophobic properties could be easily controlled with the current density and electrolyte concentration. Rough Ti substrates were also obtained in the presence of NaCl [95] and nanometer-scale dendritic structures of Zn in the presence of NaCl and NaNO$_3$ [96]. In the last example, the authors observed that the surfaces are composed of Zn, ZnO, Zn(OH)$_2$, and Zn$_5$(OH)$_8$Cl$_2 \cdot$H$_2$O. Other authors also reported the use of wire electrical discharge machining (WEDM) on aluminum and stainless steel substrates to create rough microgrooves [97, 98].

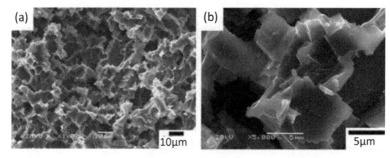

Figure 18 Aluminum substrates electrochemically machined in NaClO$_3$ and at constant current density. Reprinted with permission from ref. 94, copyright 2012, Springer.

3.1.3 Silicon

In microelectronics, semiconducting substrates such as silicon wafers are extensively used. Hence, it is very important to be able to prepare silicon wafers with various wettabilities [99–117].

Silicon substrates can be electrochemically etched in the presence of HF following the reaction Si + 6F$^-$ → SiF$_6^{2-}$ + 4e$^-$. Macroporous silicon wafers were prepared by etching in HF and at constant current density. After coating with a hydrophobic material, it was observed that the porosity percentage, the mechanical strength and the surface hydrophobicity can be controlled with the current density and HF concentration. Moreover, it was also possible to obtain superhydrophobic properties with ultra-low adhesion (θ_w = 160° and α_w < 1°) by an additional wet etching in the presence of HNO_3, NH_4F and HF, which induces a conversion from pores to pillar-like structures [99-101].

In order to obtain silicon nanowire arrays, an electroless or electrochemical etching process can be used [102-112] (Figure 19). In this process, Ag^+ ions are added in the solution. When these ions are in contact with the silicon wafer, Ag nanoparticles are deposited on the silicon following an electroless deposition (Figure 20):

$$Si + 6\,F^- + 4\,Ag^+ \rightarrow SiF_6^{2-} + 4\,Ag$$

Because Ag is more electronegative than Si, the Ag nanoparticles strongly attract electrons from Si leading to negatively charges nanoparticles, which act as catalysts for subsequent reduction of Ag^+. In this process, a SiO_2 layer is formed at the interface between Si and Ag nanoparticles, which is attacked by HF to form highly dense etch pits. At the end of the reactions, the Ag dendrites are removed using concentrated HNO_3. Lee et al. used this process and added a thermal annealing to increase the material intrinsic hydrophobicity [102]. Then, they showed that both the length of the nanowires and the surface hydrophobicity increase with the etching time. For example, for an etching time of 20 min, the nanowire length was about 13-14 µm and θ_w > 150°, but the water adhesion was very high (H_w > 48°). Moreover, these materials displayed photoluminescence and antireflectivity properties [103]. A post-treatment with a hydrophobic material can also be used to obtained superhydrophobic properties. Materials other than Ag can also be used. For example, Choi et al. deposited nanostructured Au nanostructures by glancing angle deposition before etching [109, 110]. Because silicon nanowires arrays are superhydrophilic but become superhydrophobic after post-treatment, it was also possible to prepare substrates with wettability contrasts using a mask [111, 112]. The substrates could be used for biomolecule and nanoparticle transfer.

In order to enhance the superhydrophobic properties, silicon nanowires were formed on micropattern substrates [113-117]. Lu et al. studied the nucleation and condensation of water droplet on microgrooves modified with silicon nanowire arrays [113]. They observed a preferential condensation of water vapor on the microgrooves and a continuous shedding of the dropwise condensate was observed on the surface. Moreover, the condensation also depended on the density of microgrooves. The authors also suggest high heat and mass transfer rates on these surfaces. In order to modify the adhesion properties, Lee et al. coated silicon nanowire arrays by palladium [114]. They observed a reversible wettability switching between superhydrophobic (low adhesion) to parahydrophobic (high adhesion) when the substrates are in presence of air and H_2, respectively. Moreover, the authors showed that these substrates with switchable adhesion properties can potentiate the therapeutic efficacy of 3D stem cell spheroids [115]. Inspired by the air-retaining capability of superhydrophobic submerged insects, Feng et al. also reported the preparation of electrodes by depositing platinum on silicon nanowire arrays [116]. Indeed, sufficient and constant oxygen is necessary for enzymatic reaction interface. Here, the authors showed that the presence of nanowires enhances the detection of glucose.

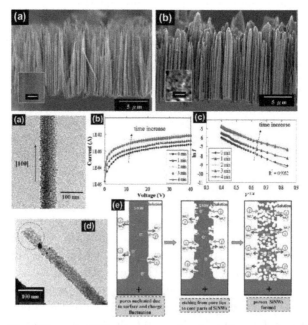

Figure 19 SEM and TEM images and schematic representation of porous silicon nanowire arrays produced by an electrochemical etching process. Reprinted with permission from ref. 103, copyright 2012, Elsevier.

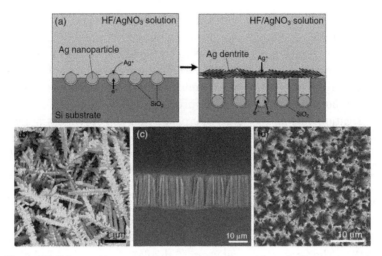

Figure 20 Schematic representation of the formation of silicon nanowire arrays produced by an electroless etching. Reprinted with permission from ref. 102, copyright 2013, Elsevier.

3.2 Plasma Processes

3.2.1 Plasma Etching

In a plasma chamber, a substrate is put in contact with ionized species produced by creating an electric field between two electrodes. The interaction of the plasma with the substrate can have different effects such as the surface cleaning, the formation of chemical groups or the formation of surface structures. These effects are highly dependent on the plasma parameters such as the used gas, the pressure, or the power [118]. For example, nanostructures were reported on carbon nanofibers using an O_2 plasma treatments. Usually, the plasma treatments are especially performed on polymer substrates, which are more sensitive to temperature. However, it is also possible to etch silicon substrate using fluorinated gases such as SF_6. Bøggild et al. developed silicon nanograss by an anisotropic reactive ion etching (RIE) system using SF_6 and O_2 as gases [119]. The authors reported the possibility of obtaining overhanging nanostructures by increasing the O_2 flow with enhanced oleophobic properties. These substrates could also be used for electrowetting experiments after deposition of a thin high k-dielectric HfO_2 layer [120].

Deep Reactive Ion Etching (DRIE) processes were also used on SiC substrates to produce nanostructures [121, 122]. Nanopillar, nanocone, and nanowhisker arrays were also reported on diamond and boron nitride (BN) substrates [123, 124]. In order to develop both silicon microcones and nanograss, Rühe et al reported a cryogenic DRIE process ($T = -105°C$) in the over passivation regime to create microcones on silicon wafers before the formation of silicon nanograss [125] (Figure 21). After post-treatment, superhydrophobic properties were obtained. Choi et al. reported the formation of silicon cylindrical nanoshell arrays using a novel sublithographic patterning ("spacer lithography") [126] (Figure 22). Here, a sacrificial oxide layer is deposited, SiO_2 pillars are obtained by photolithography and RIE, and after the oxide pillar inside the nanoshells are removed by buffered oxide etchant (BOE). The nanoshell substrates displayed superhydrophobic properties with $\theta_w = 166°$ and $H_w = 5°$ without post-treatment.

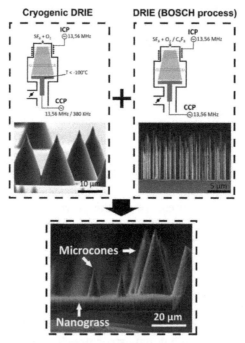

Figure 21 Microcones and nanograss developed on silicon wafers by a cryogenic and a classic DRIE process in the presence of SF_6 and O_2 gases. Reprinted with permission from ref. 125, copyright 2014, American Chemical Society.

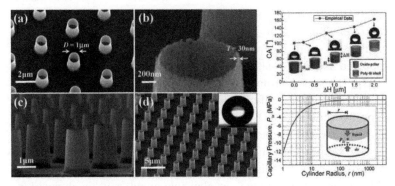

Figure 22 Silicon cylindrical nanoshell arrays obtained using a novel sublithographic patterning ("spacer lithography"). Reprinted with permission from ref. 126, copyright 2010, American Chemical Society.

3.2.2 Sputter Deposition

The sputter deposition is a physical vapor deposition process allowing the ejection of a material (target) onto a substrate. Radio frequency (RF) magnetron sputtering is a method used in the literature to fabricated superhydrophobic nanostructures metals and metal oxides (Figure 23). The structure formation can be controlled with different parameters such as the gas flow, the flow rate and the temperature. For example, aluminum, copper, silver, gold, zinc, and cadmium nanocrystals with superhydrophobic properties were reported with this technique [127–130]. For example, gold nanostructures were sputtered on copper meshes substrates [131]. After post-treatment with a thiol mixture of HS(CH$_2$)$_9$CH$_3$ and HS(CH$_2$)$_{10}$COOH, the meshes should be used to separate oil/water mixture bidirectionally.

Figure 23 ZnO particles deposited by RF magnetron sputtering. Reprinted with permission from ref. 129, copyright 2015, Elsevier.

Different compounds can also be deposited with this process. Gonzalez-Elipe et al. deposited Ag@TiO$_2$ core@shell nanorod arrays with superhydrophobic properties [132]. Yamashita et al. reported nanocomposites of TiO$_2$ and PTFE with a rod-like structure [133]. The nanocomposite substrate displayed superhydrophobic properties with self-cleaning and photocatalytic properties. Superhydrophobic TiO$_2$-Cu$_2$O coatings were also reported using phase-separated magnetron sputtering technique [134].

By simply inclining the substrate inclination during sputter deposition process, it is possible to obtain inclined nanostructures. This technique is known as glancing angle sputter deposition. For example, Singh et al. prepared inclined silver nanorod arrays using an inclination angle of 85° [135–137]. The average diameter and number density of nanorods could be easily controlled with the deposition rate or time. The substrates were superhydrophobic with θ_w = 157° and α_w = 5° without post-treatment and could be used as surface enhanced fluorescence spectroscopy (SEFS). Moreover, inclined silver nanorods were deposited on wrinkled PDMS leading to anisotropic wettability [137]. The droplet could move easily but only along the direction parallel to the wrinkles. The wettability properties could be controlled with the mechanic strain, which induces changes in amplitude and periodicity of the wrinkles. These substrates are extremely interesting for the control of microliter water droplets in a preset direction [138]. Indeed, the substrates exhibited more than threefold fluorescence signal enhancement than conventional silver nanorods films. Magnesium nanorod arrays coated with a PTFE shell were also reported using a two-step sputtering process [139] (Figure 24). The presence of PTFE allows obtaining superhydrophobic properties with a long-term storage stability. Moreover, the materials displayed excellent energetic capacity and could be used for microactuation/micropropulsion. CaF$_2$ nanorod arrays also displayed self-cleaning and antireflection properties [140]. The authors showed that these materials can be used in organic photovoltaic cells with excellent power conversion capacity. Lin et al. also used SiO$_2$ nanorods arrays on photonic crystals for achieving maximal solar absorption [141].

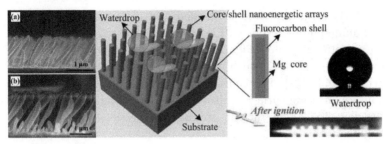

Figure 24 Inclined Mg nanorods coated with a PTFE shell deposited by glancing angle sputter deposition for energy systems. Reprinted with permission from ref. 139, copyright 2014, American Chemical Society.

The surface morphology of the nanostructures can also be changed using various strategies. Veinot et al. showed that various surface structures such as vertical, inclined, and also helicoidal nanostructures can be obtained by inducing a substrate rotation during the process [142, 143] (Figure 25). Moreover, the authors observed that the vertical nanostructures possess low water adhesion, while the inclined and helicoidal nanostructures possess high water adhesion.

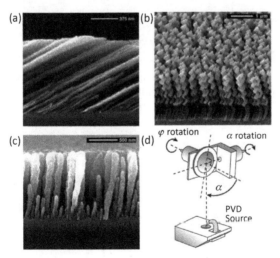

Figure 25 Inclined and helicoidal SiO_2 nanostructures by rotation during a glancing angle sputter deposition process. Reprinted with permission from ref. 143, copyright 2004, American Chemical Society.

Biris et al. prepared controlled and self-organized tungsten and aluminum nanorods with a hexagonal geometry by deposition on smooth substrates (Figure 26) but also on anodized aluminum oxides (Figure 27), which have hexagonally packed nanopores arrays [144–147]. The characteristics of the nanorods and as a consequence the surface hydrophobicity could be controlled with the deposition time, the gas pressure or the substrate inclination angle.

Figure 26 Tungsten triangular nanorods obtained by glancing angle sputter deposition with different substrate inclination and Ar pressure. Reprinted with permission from ref. 144, copyright 2011, American Chemical Society.

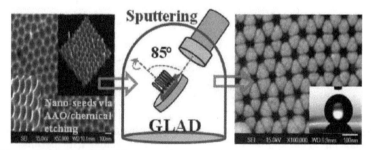

Figure 27 Self-organized hexagonal arrays of tungsten triangular nanorods obtained by glancing angle sputter deposition on anodized aluminum oxide (AAO) membranes. Reprinted with permission from ref. 145, copyright 2011, American Chemical Society.

Other techniques were also used in the literature. The atomic layer deposition was used to deposited inorganic materials such as Al_2O_3 [148, 149], TiO_2 [150], ZnO [151] as well as rare earth oxides [152]. Ion implantation was also used on InSb and GaSb substrates to obtain nanoporous superhydrophobic substrates with antireflective properties [153, 154].

3.3 Laser

The application of laser of specific wavelength and frequency can etch diverse substrates, even metallic ones [155, 156]. By laser treatment, it is possible to create microlines or microgrooves by scanning the laser in one dimension and micropillars by scanning in two dimensions [157-170]. The surface topography can be controlled with many parameters such as used laser, the laser fluence, the scanning speed, the scanning interval, and the scanning number on the surface structures. For example, Zhou et al. studied their influence on silicon substrates etched to form microgrooves with femtosecond and nanosecond laser [157]. They observed that the microgrooves fabricated with femtosecond lasers are smoother and with smaller top width of groove. However, by overlapped etching for many times, both micro- and nanoscale structures are observed with higher surface hydrophobicity (Figure 28).

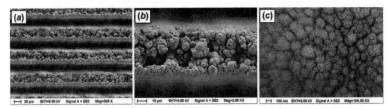

Figure 28 Rough microgrooves obtained by nanosecond laser treatment of aluminum substrates by scanning the laser in one dimension. Reprinted with permission from ref. 159, copyright 2015, American Chemical Society.

Many other authors also reported the formation of both microgrooves and nanostructures on copper, aluminum, glass, or graphene oxide substrates by using nanosecond laser [158-166]. In order to obtain highly rough and porous microgrooves on silicon substrate, Zhao et al. used a chemical etching in HF after the laser treatment. The θ_w values of about 150° were measured on the surfaces [167]. Sun et al. deposited nanostructured graphene layers after laser irradiation [168]. The resulting substrates displayed both superhydrophobic and iridescence properties. By contrast, other authors used laser on anodized titania nanotubes or nanorod arrays in order to form lines made of TiO$_2$ nanotubes or nanorods [169, 170].

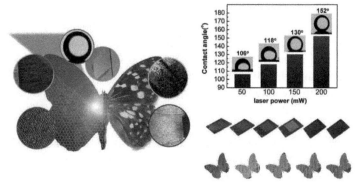

Figure 29 Superhydrophobic and iridescence properties obtained by deposition of graphene layers on laser-treated substrates. Reprinted with permission from ref. 168, copyright 2016, Wiley.

Periodic micropatterned copper substrates with micro-protrusions and micro-pits were developed by using femtosecond laser by scanning in two dimensions [171–177]. The topography of the microstructures could be easily controlled with the scanning speed of the laser beam [171–174]. More precisely, the authors observed the substrates with deep microstructures obtained with the lowest scanning speed (10 mm s^{-1}) showed superhydrophobic properties with extremely low water adhesion, while the water adhesion increased when the surface microstructures were flat. The same authors also studied the influence of the laser fluence, the scanning speed the scanning interval, and the scanning number on the surface structures. They observed the formation of nanoscale structures on the micro-patterned substrates when the laser fluence is high or the laser scanning speed is low. The formation of both nanostructures and micropatterns was found to be necessary to obtain stable and robust superhydrophobic properties. Moreover, by combining a femtosecond laser system with a computer, it was possible to design different patterns with various wetting properties. Kietzig et al. developed regular posts arranged in a square pattern and rhombic posts arranged in a hexagonal pattern and studied the drag reduction on these substrates [177] (Figure 30). They observed that even if the rhombic posts facilitate the Cassie–Baxter state, the higher slip length was measured on the regular posts.

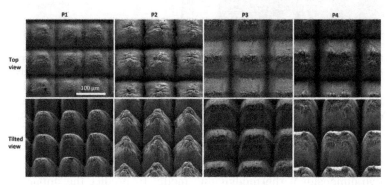

Figure 30 Regular posts arranged in a square pattern and rhombic posts arranged in a hexagonal pattern obtained by laser treatment of copper substrates and by scanning the laser in two dimensions with different laser parameters. Reprinted with permission from ref. 177, copyright 2016, Royal Society of Chemistry.

Micropatterned substrates with mesh-like porous structures were also observed on silicon, titanium, and zinc substrates by using femtosecond laser [178–187]. Superhydrophobic properties (θ_w = 158° and α_w = 4°) were observed on silicon substrates [179]. By contrast, on titanium substrates and zinc substrates, the presence of TiO_2 and ZnO was observed, respectively, leading to switchable superhydrophobic/superhydrophilic properties by alternating UV treatment and dark storage [178–180] (Figure 31). Aluminum substrates with crater-like pits were reported by Liu et al [181]. After etching in HNO_3 and $Cu(NO_3)_2$, nanoporous substrates were obtained with extremely high water adhesion.

Picosecond and nanosecond lasers were also used in the literature [188–191]. For example, Jagdheesh studied the use of picosecond lasers on alumina substrates [188]. Square micropillars with low inter-plot distance were developed using different number of pulses. Nanoroughness was also observed on the microstructures.

30 | Fabrication Processes

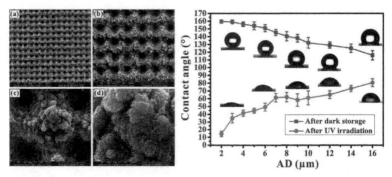

Figure 31 Micropatterned substrates with mesh-like porous structures by laser treatment of titanium substrates. The resulting substrates also displayed switchable superhydrophobic/superhydrophilic properties. Reprinted with permission from ref. 178, copyright 2015, Royal Society of Chemistry.

Stainless steel substrates were also studied by laser treatments [192–203] (Figure 32). Hatzikiriakos et al. used a two-dimensional thermodynamic model to predict the apparent contact angle and contact angle hysteresis of sinusoidal and parabolic patterned obtained by laser treatments [192–194]. They showed that the ratio of the height to base diameter is a crucial parameter to obtain superhydrophobic properties. Moreover, the parabolic structures gave high water adhesion, while the sinusoidal one low water adhesion. The authors also studied the influence of laser parameters such as the laser fluence, and the scanning speed nanoripples were observed on the microstructures [194].

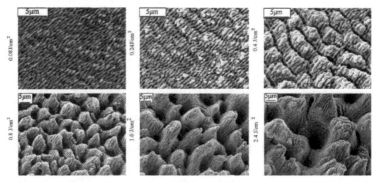

Figure 32 Nanostructured parabolic structures obtained by laser treatment of stainless steel substrates. Reprinted with permission from ref. 195, copyright 2009, Elsevier.

The presence of the nanoripples is also fundamental to lead to superhydrophobic properties. Moreover, the structural color, which is directly dependent on the periodic patterns, can also be controlled with the experimental conditions [198, 199]. Similar structures were also obtained on polycrystalline $Ni_{60}Nb_{40}$ [200].

Mazumder et al. reported the fabrication of microcones arrays on stainless steel substrates using a femtosecond laser treatment [201] (Figure 33). They observed the formation of microcones using low pulse energy (< 10 µJ) and high repetition rate (≥ 500 kHz). Superhydrophobic properties were observed after post-treatment with a fluorinated silane. Micropillar arrays with re-entrant structures were fabricated by laser treatment, insulating, mechanical polishing and electrodeposition [202].

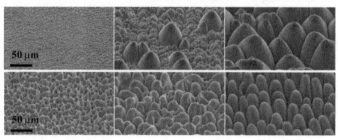

Figure 33 Microcone arrays obtained by laser treatment of stainless steel substrates using low pulse energy and high repetition rate. Reprinted with permission from ref. 201, copyright 2012, IOP Publishing.

Fotakis et al. also reported a strategy to fabricate conical spikes forests on various substrates such as silicon [204–207]. Here, the authors used femtosecond lasers but at low pressure (10^{-2} mmbar) and in the presence of SF_6 gas (Figure 34). The height of the spikes could be controlled with the laser fluence. After post-treatment, the substrates displayed superhydrophobic properties with θ_w of about 154° and H_w = 5°. By coating the substrates with a thermally grown silicon oxide layer, the substrates could be used for electrowetting experiments [206]. After coating with ZnO, it was also possible to obtain substrates with reversible wettability from superhydrophobic to superhydrophilic by UV irradiation and heating or dark storage. Titanium nanocone arrays were also reported by femtolaser treatment at low pressure (1 mbar) [208, 209]. The formation of cone and hole structures was also reported by direct laser interference manufacturing.

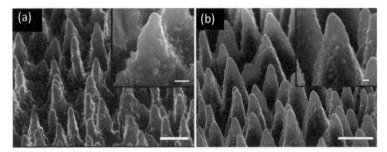

Figure 34 Microspike arrays obtained by laser treatment of silicon substrates at low pressure and in the presence of SF$_6$ gas. Reprinted with permission from ref. 204, copyright 2008, Springer.

3.4 Etching in Basic Media

When a non-noble metal substrate is put in contact with oxygen (O$_2$), it can react following an oxido-reduction reaction: 2 M + O$_2$ → 2 M$^+$ + 2 O^{2-} → 2 MO. As a function of the metal and the temperature, different nanostructures can be formed on the substrate [210–214]. Rough titanium substrates made of nanometer-scale TiO$_2$ were prepared by thermal oxidation at 900–1100°C [213]. (θ_w = 166° and α_w = 2°) were obtained after post-treatment. Superhydrophobic ZnO nanowire arrays were also grown on zinc substrates by thermal oxidation at 500–600°C.

A non-noble metal substrate can also be oxidized in water: M + H$_2$O → MO + H$_2$. Different crystalline structures can be obtained as a function of the temperature and the immersion time. For example, the immersion of aluminum substrates in boiling water led to the formation of boehmite (α·Al$_2$O$_3$·H$_2$O) petal-like crystalline nanostructures: 2 Al + (3 + x) H$_2$O → Al$_2$O$_3$·xH$_2$O + 3 H$_2$ [210, 211]. After post-treatment, superhydrophobic substrates (θ_w = 163°) were obtained with anticorrosion properties. The surface morphology could also be modified by adding chemical compounds such as dimethylformamide (DMF).

3.4.1 Copper

When a nonmetal substrate is immersed in a basic aqueous solution, many hydroxides and oxides can be formed as a function of the metal, the solution pH, and subsequent heat treatment to remove aqueous molecules. The reaction is

$$4\,M + O_2 + 2\,H_2O \rightarrow 4\,M^+ + 4\,OH^-.$$

Because of its extremely high sensitivity to water and the possibility of obtaining various surface morphologies, copper has been the most studied material in the literature. Many authors reported the possibility of obtaining nanometric CuO/Cu(OH)$_2$ nanostructures by oxidation of copper substrates using a persulfate or H$_2$O$_2$ as an oxidizing agent and KOH or NaOH as a base [215–221]. The suggested chemical reactions are

$$Cu + 2\,OH^- + S_2O_8^{2-} \rightarrow Cu(OH)_2 + 2\,SO_4^{2-}$$

$$Cu(OH)_2 + 2\,OH^- \rightarrow Cu(OH)_4^{2-} \rightleftharpoons 2\,CuO + 2\,OH^- + H_2O.$$

Using this process, different nanostructures such as nanowires, nanorods, nanotubes, and nanoflowers were reported to display superhydrophobic properties after post-treatment (Figure 35).

Figure 35 CuO nanoflowers and nanograss obtained by immersing Cu substrates in NaOH and (NH$_4$)$_2$S$_2$O$_8$ [218].

The structures were also formed on copper mesh substrates and copper foams for applications in oil/water separation [222–227]. Inspired by cacti, Dai et al. fabricated a novel centrifugation-assisted fog-collecting device using copper foams modified with copper oxide nanoneedles [227] (Figure 36). The superhydrophobic copper foams showed exceptional collection efficiency (86% for a rotation of 1500 rpm).

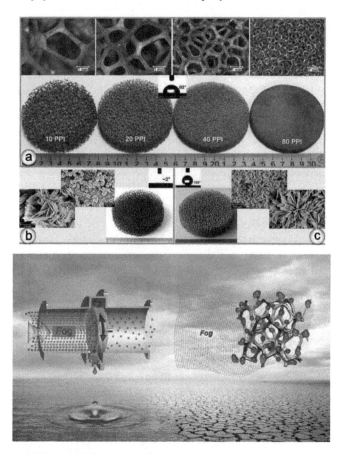

Figure 36 CuO nanoflowers and nanograss obtained by immersing Cu foams in NaOH and $K_2S_2O_8$ and their use for novel centrifugation-assisted fog-collecting devices Reprinted with permission from ref. 227, copyright 2016, American Chemical Society.

In order to develop superoleophobic properties on polymer substrates, $Cu(OH)_2$ nanowire array were also grafted on polymeric substrate using dopamine [228] (Figure 37). Then, an aluminum nanolayer was deposited on the $Cu(OH)_2$ nanowires and the resulting substrates were treated by boiling water to form petal-like $\alpha \cdot Al_2O_3 \cdot H_2O$ nanostructures. After fluorination, the substrates could repel many kinds of liquid even decane.

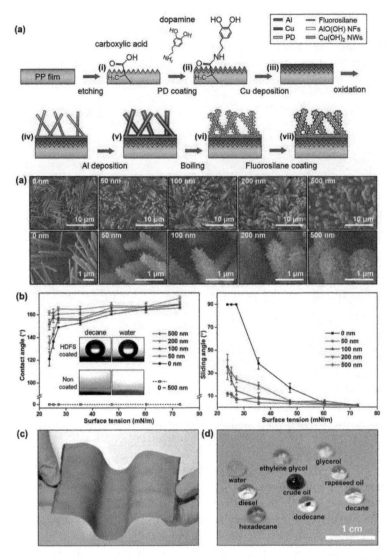

Figure 37 Superoleophobic polymer substrates by grafting $Cu(OH)_2$ nanowire arrays with dopamine, modification with an aluminum nanolayer and treatment by boiling water to form petal-like $\alpha \cdot Al_2O_3 \cdot H_2O$ nanostructures and the $Cu(OH)_2$ nanowires. Reprinted with permission from ref. 228, copyright 2016, Royal Society of Chemistry.

By using $NaHCO_3$ and $K_2S_2O_8$, superhydrophobic $Cu_2(OH)CO_3$ nanorods arrays were also obtained [229]. The crystalline structures could be modified by replacing KOH or NaOH by other bases such as H_2NNH_2 or $H_2N(CH_2)_6NH_2$ or even bases such as NH_3 able to form complexes with metal ions (for instance, $Cu[NH_3]_n^{2+}$) [230–245]. For example, sisal-like nanoribbons were observed in the presence of NH_4OH [237] (Figure 38).

Figure 38 CuO sisal-like nanoribbons and nanosheets obtained by immersing Cu substrates in KOH, NH_4OH and $K_2S_2O_8$. Reprinted with permission from ref. 237, copyright 2012, American Chemical Society.

Using NH_3 vapor, $Cu(OH)_2$ ball-like structures composed of nanoslices were observed [238] (Figure 39). Moreover, the water adhesion could be adjusted from 65 to 14 μN on these substrates by using alkyl thiols of various alkyl chain lengths. The crystalline structures could also be changed by adding a copper source in the solution. Superhydrophobic flower-like [$CuSO_4 \cdot 3Cu(OH)_2 \cdot H_2O$] and dandelion-like $Cu_4SO_4(OH)_6$ structures were observed after the immersion of copper substrates in $CuSO_4$ aqueous solution [240–242]. Slice-like $Cu_2(OH)_3NO_3$ structures were also obtained by the immersion of copper substrates in a solution of NaOH and $Cu(NO_3)_2$ [243].

In order to develop multilevel hierarchical nanostructures with a high degree of re-entrant curvatures from CuO nanowires, the deposition of hydrocarbon or fluorocarbon wax crystals on these nanowires was reported in the literature by thermal evaporation [246–247]. The surface properties could be controlled by tuning the CuO nanowires, changing the used wax or its amount. The best results were obtained with fluorocarbon wax crystals with $\theta_{hexadecane} \approx 150°$ and $H_{hexadecane} < 10°$.

Figure 39 Cu(OH)$_2$ ball-like structures obtained by immersing copper substrates NH$_3$ vapor. Reprinted with permission from ref. 238, copyright 2013, Royal Society of Chemistry.

3.4.2 Others

Other materials were also used in the literature. Titania nanowire arrays were created by immersing titania substrates in the presence of melamine (C$_3$H$_6$N$_6$), H$_2$O$_2$ and HNO$_3$. Cheng et al. used this method to modify titanium meshes and the resulting meshes could be used for oil/water separation [248].

Al(OH)$_3$ nanoflakes and flower-like structures were observed by immersing aluminum substrates in NaOH or H$_2$N(CH$_2$)$_6$NH$_2$ [249–251]. Triangular prism arrays were obtained by immersion in NH₄OH at 80°C [252] (Figure 40). After post-treatments, superoleophobic properties ($\theta_{hexadecnae}$ = 151° and $\alpha_{hexadecane}$ = 25°) were observed. Dandelion-like ZnO microspheres composed of nanoneedles and ZnO sub-micronic rod arrays were also reported using NH₄OH and H₂N(CH₂)₂NH₂, respectively [253, 254].

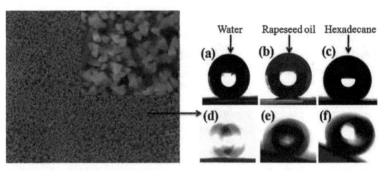

Figure 40 Al(OH)₃ triangular prism arrays obtained by immersing aluminum substrates in NH₄OH at 80°C. Reprinted with permission from ref. 252, copyright 2015, Elsevier.

3.5 Anodization

3.5.1 Aluminum

Anodization consists of the artificial formation of an oxide layer by electrochemical oxidation of a non-noble metal substrate. This is a very-known and used method in order to create structured oxide layers on metal substrate.

In the case of aluminum, the reaction at the anode is

$Al \rightarrow Al^{3+} + 3e^-$

$2Al^{3+} + 3H_2O \rightarrow Al_2O_3 + 6H^+$.

In this reaction, the oxidation of Al into Al₂O₃ and the dissolution of Al₂O₃ by H⁺ are in competition. The characteristics of the surface structures are highly dependent on electrochemical parameters such as the nature of the acid used in the process (phosphoric acid, sulfuric acid, or oxalic acid, for example), the electrochemical method (constant potential and constant intensity, for example) and their parameters, or the distance between the working electrode and the counter-electrode [255] (Figure 41).

The conventional anodization (in H_2SO_4) of aluminum substrate is known to produce on the surface hexagonally packed nanopores arrays. For example, Tsujii was the first to report the use of anodization of aluminum substrates to produce superoleophobic properties [256, 257]. The substrates were obtained after 3 h in 0.5 M H_2SO_4 and using a current density of 10 A/cm^2 leading to rough surfaces with fractal structures. After treating with a perfluorinated phosphate, the substrates displayed superoleophobic properties with $\theta_{hexadecane}$ = 135.5°.

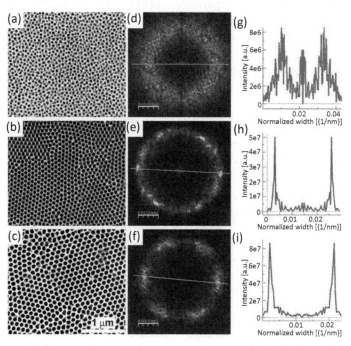

Figure 41 Hexagonally packed nanopores arrays obtained by anodizing aluminum substrates in oxalic acid and using different voltages. Reprinted with permission from ref. 268, copyright 2015, Elsevier.

In order to enhance the surface hydrophobicity and oleophobicity, the nanopores can be widen to induce the growth of nanopillars from the nanopores using mild (in H_3PO_4, for example) or hard (in oxalic acid, for example) anodization processes or even two-step anodization (Figure 42). The pore widening can also be performed by chemical etching in H_3PO_4, oxalic acid or H_2SO_4, for example [258–273]. For example, Zhang et al. studied the anodization in H_3PO_4 as a function of the anodization time [260]. After 1 min anodization, hexagonally packed nanopores arrays are observed with pore diameter of 100 nm and θ_w = 104.6° with strong water adhesion. Then, the pore diameter increases as the anodization time while the boundary between the pores becomes thinner. For an anodization time of 20 min, the presence of nanopillars on the nanopores was observed and θ_w increased to 167.1° with low water adhesion. The substrates also displayed drag reduction [270], anticorrosion properties, high transparency [271] and maintain their superhydrophobic properties even after storage in air on water for 3 months. Superoleophobic properties with $\theta_{hexadecane}$ = 153.2° and $\alpha_{hexadecane}$ = 3° were also obtained after post-treatment with a perfluorinated phosphate [273].

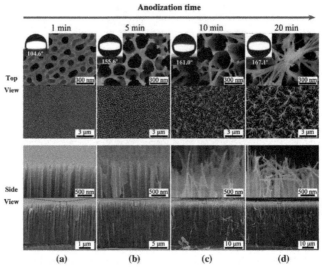

Figure 42 Anodized aluminum substrates in H_3PO_4 and using different anodization time: growth of nanopillars on nanopores. Reprinted with permission from ref. 260, copyright 2015, Elsevier.

In order to enhance the surface properties, aluminum substrates were first etched in HCl to form microporous "building block" architectures before to induce nanopores on the microstructures by anodization [274, 275]. The resulting substrates displayed superhydrophobic and anti-icing properties. Aluminum meshes were also etched and anodized to obtain both micro- and nanostructures [276] (Figure 43). After posttreatment with an alkylphosphate and a fluorinated phosphate, the meshes showed superoleophilic and superoleophobic properties, respectively. These materials are excellent candidates for oil/water separation. The surface properties can also be enhanced by posttreatment on anodized aluminum oxide (AAO) substrates [277–283]. For example, after depositing Ag nanoparticles, substrates displaying superoleophobic ($\theta_{hexadecane}$ = 155° and $\alpha_{hexadecane}$ = 5°) and antibacterial properties were reported [278]. Wang et al. induced the growth of Co_3O_4 nanosheets by hydrothermal process [279]. The resulting substrates displayed superoleophobic properties and also slippery properties interesting for anti-icing properties. Bengu et al. induced the growth of carbon nanotubes over AAO substrates [283]. Porous structures of different sizes were obtained using different electrolytes (H_2SO_4, H_3PO_4 and oxalic acid). After depositing Co-Al catalysts, carbon nanotubes were created by CVD. Here, the highest superhydrophobic properties were observed with vertically aligned carbon nanotubes.

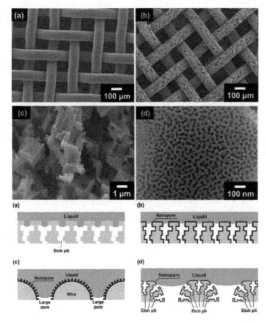

Figure 43 Superoleophobic aluminum meshes after chemical etching in HCl and $CuCl_2$ and a two-step anodization in H_2SO_4 and H_3PO_4. Reprinted with permission from ref. 276, copyright 2016, American Chemical Society.

By immersing aluminum oxide substrates in solution containing metal ions such as Mg^{2+}, Co^{2+}, Ni^{2+}, and Zn^{2+}, it was possible to obtain layered double hydroxides (LDH) with nanosheet morphology [284]. Superhydrophobic properties were obtained after post-treatment with a fluorinated silane. It was also possible to incorporate inside these LDH different anions such as CO_3^{2-} and NO_3^- or even dodecanoate to lead to superhydrophobic properties [285–287]. Moreover, the authors also obtained UV-switchable materials by incorporation of fluorinated substituted azobenzene anions (7-[(trifluoromethoxyphenylazo)phenoxy] pentanoate). The surface wettability could be switched from superhydrophobic to hydrophilic by UV radiations and from hydrophilic to superhydrophobic by visible light irradiation.

Anodic aluminum oxide layers can be released from their substrates and used as nanoporous membranes. For example, in order to produce transparent and flexible materials, the group of Kim transferred the pillar structures onto a polymer substrate (PET) by bonding the AAO membrane using a sacrificial photoresist layer [288] (Figure 44). Then, the removal of the sacrificial layer in acetone and chemical etching in $CuCl_2$ and HCl induces pore widening and collapsed nanopillars. The substrates were also superhydrophobic with low water adhesion (θ_w = 164.4°, H_w = 5.2° and $\alpha_w \approx 10°$). The same procedure was also used to transfer these nanostructures on the top of the pillars of micro-pattern PET substrates [289].

Neto et al. used AAO membranes as a template for the electrodeposition of nickel and platinum [290]. After membrane dissolution, the formed dense and aligned nickel nanowires displayed after post-treatment superhydrophobic properties with low adhesion (θ_w = 154° and H_w = 12°). Here, the aligned nickel nanowires had a high aspect ratio ($\approx$250) with a $\varnothing$ of 200 nm and a length of several tens of μm. By contrast, the platinum nanowires were more disordered, with a wave-like appearance and displaying higher water adhesion. Conducting polymers can also be electrodeposited on AAO membranes [291]. Jiang et al. reported the fabrication of polypyrrole/Fe_3O_4 nanotube arrays by electrodeposition of polypyrrole in the presence of Fe_3O_4 nanoparticles with electromagnetic properties. The study of the growth of polyaniline on AAO membranes was also reported in the literature with the formation of superhydrophobic and conductive membranes [292, 293]. Zhu et al. also electrodeposited a conductive salt ([(CH$_3$)$_4$N][Ni(C$_3$S$_5$$^{2-}$)$_2$]), except that the membrane was not dissolved after deposition. A rose-like morphology was obtained with superhydrophobic properties (θ_w = 152° and α_w = 2°) [294].

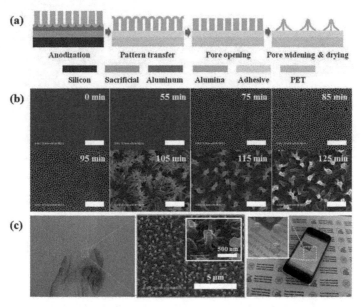

Figure 44 Highly flexible and transparent superhydrophobic films prepared by transfer from an anodized aluminum oxide membrane. Reprinted with permission from ref. 288, copyright 2014, Royal Society of Chemistry.

It is possible to fill in the nanopores of AAO by drop casting a polymer dissolved in organic solvent to induce the polymer diffusion into the membrane nanopores [295] (Figure 45). It is also possible to induce polymerization inside the nanopores. After peeling off, a polymer material with the inverse nanostructures of AAO membranes can be formed. Using this technique, Bayindir et al. developed polycarbonate with vertically aligned nanoplots. Moreover, the ⌀ and height of the nanoplots could be easily controlled with the nanoporosity of the AAO membranes. The substrates displayed antireflectivity, surface-enhanced Raman scattering (SERS) features, and highly hydrophobic properties (θ_w = 145.4°).

Figure 45 Nanostructured superhydrophobic polymers by casting inside an anodized aluminum oxide membrane. Reprinted with permission from ref. 295, copyright 2013, Royal Society of Chemistry.

Polymer materials can also fill nanopores by extrusion at a relatively high temperature above the polymer T_g [296–301]. For example, Zhang et al. created superhydrophobic materials by the extrusion of high-density polyethylene inside AAO membranes [297]. Here, the authors tuned the diameter of the nanopore membranes and the extrusion pressure in order to control the nanofiber size and the resulting superhydrophobic properties.

By applying an AAO membrane of substrate coated with melted polystyrene layers, membrane dissolution, polystyrene nanoplots, or nanotubes arrays were reported in the literature as a function of the application time [298]. The substrates displayed extremely high water adhesion and the with a possible control with the tip geometry.

Different strategies were also used in the literature in order to completely replicate AAO nanostructures on other substrate [299–311]. For example, Choi et al. transferred h-PDMS hairy nanostructures on glass substrates [308, 310]. The substrates displayed extremely high θ_w = 150.5° and high water adhesion. Deng et al. also transfer the porous structures of AAO membranes on poly(methyl methacrylate) (PMMA) substrates.

3.5.2 Titanium

The anodization process is highly dependent on the metal substrate. Hence, depending on the metal substrate, the electrolyte, and the anodizing parameters, different nanostructures can be obtained. Mozalev et al. studied the anodization of aluminum-on-niobium (Al/Nb) and aluminum-on tantalum (Al/Ta) layers in citric acid and at high voltage (400–480 V) [312] (Figure 46). They created bowl-like and mushroom-like structures, respectively, with superhydrophobic properties (θ_w = 156–158°).

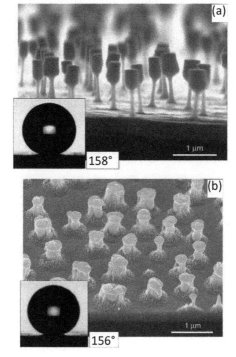

Figure 46 Bowl-like and mushroom-like structures obtained by anodizing aluminum-on-niobium (Al/Nb) and aluminum-on tantalum (Al/Ta) layers in citric acid. Reprinted with permission from ref. 260, copyright 2012, Elsevier.

The anodization is extremely used in the literature with titanium and specially to form TiO_2 nanotube arrays [313–325]. For titanium, it is necessary to use F^- ions for anodization. The reaction is Ti + 2H$_2$O → TiO$_2$ + 4H$^+$ + 4e$^-$ but in the presence of F⁻ ions in the electrolyte, the formation of nanopits and the resulting nanopores is induced by the following reaction $TiO_2 + 6 F^- + 4H^+ \rightarrow TiF_6^{2-} + 2H_2O$. The ∅ and the height of the nanopores can be easily controlled with the electrolyte (aqueous or nonaqueous), the anodization potential and time, the pH or the temperature. For example, Lai et al. developed arrays of TiO_2 nanotubes (∅ ≈ 100 nm and length ≈ 400 nm) by anodizing in 0.5 wt% HF electrolyte at 20 V for 20 min and finally annealed at 450°C in air for 2 h to obtain anatase form [314–317] (Figure 47). Due to photocatalytic property of TiO_2, reverse wettability pattern could be realized by UV illumination. After post-treatment with a perfluorinated silane, superhydrophobic properties were obtained (θ_w = 160°). Moreover, substrates with wettability contrasts (superhydrophobic/superhydrophobic) could be prepared through self-assembly and photocatalytic lithography, for example. They could act as a 2D scaffold for site-selective cell immobilization and reversible protein absorption and a wide range of applications can be envisaged such as for biomedical devices (high-throughput molecular sensing, targeted antibacterials, drug delivery).

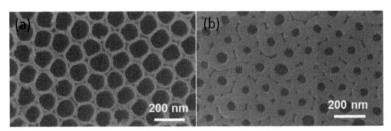

Figure 47 Nanotubes obtained by anodizing titanium substrates in NH_4F at low voltage and annealing. Reprinted with permission from ref. 318, copyright 2014, Elsevier.

Superoleophobic properties were prepared by first etching titanium substrates to obtain microstructures and anodization to develop nanotubes [321]. The nanotube wall thickness was less than 10 nm and the pore diameter 95–105 nm leading to superoleophobic properties with $\theta_{\text{olive oil}}$ = 151°. In order to prepare anisotropic superoleophobic surfaces, micro- and nanostructured TiO_2 surfaces were produced by laser machining (patterned lines with various depths) followed by anodization [316]. The highest properties along the lines were $\theta_{\text{hexadecane}}$ = 157° and

$\alpha_{\text{hexadecane}}$ = 3.5°. The oil adhesion could be controlled with the distance of the patterned lines the density of nanotubes but also by UV illumination and annealing.

Other nanostructures can be obtained by titanium anodization [323–325]. For example, pinecone-like anatase microstructures were obtained by anodization with NH_4F at high voltage ≈ 50 V and without annealing [324, 325] (Figure 48). The structures could be tuned with the anodization time and potential. After post-treatment, the substrates displayed superhydrophobic properties (θ_w = 161.4° and α_w ≈ 0°), superoleophobicity, long-term stability, mechanical robustness, and anticorrosion properties.

Figure 48 Pinecone-like microstructures obtained by anodizing titanium substrates in NH_4F at high voltage and without annealing. Reprinted with permission from ref. 325, copyright 2016, Elsevier.

3.5.3 Copper

Copper oxide and hydroxide nanoneedles were also reported by anodization of copper substrate in KOH or NaOH and at constant current density [326–329]. The surface morphology could be controlled with the current density, the anodization time. The post-treatment with alkyne thiol could lead to superhydrophobic properties but also changed the surface morphology due to the formation of $Cu(SC_nH_{2n+1})_2$ complexes. Copper oxide nanoneedles were prepared on copper mesh substrates [330–331]. After fluorination, the resulting meshes could be used in rolling-spheronization granulation (Figure 49).

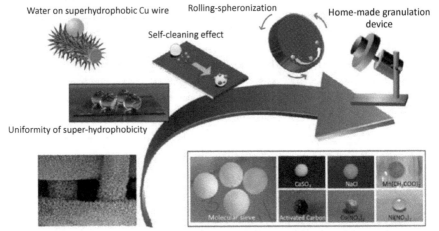

Figure 49 Anodized copper meshes using in rolling-spheronization granulation. Reprinted with permission from ref. 331, copyright 2016, American Chemical Society.

3.5.4 Others

Zinc substrates were also used in the literature for anodization. Zheng et al. developed various ZnO nanostructures including nanodots, nanowires and nanoflowers using mixtures of water, HF and methanol [332]. The authors tuned the surface morphology with HF concentration, methanol volume fraction, anodization time and the applied voltage. The substrates displayed highly hydrophobic properties, photoluminescence (near UV absorbance peak at 380 nm and very strong green emission peak at 576.2 nm), and electrowetting properties. Mattia et al. reported also various ZnO nanostructures as a function of the electrolyte (H_3PO_4, HNO_3, HCl, oxalic acid, NaOH, $KHCO_3$), potential, temperature and anodization time [333]. Li et al. studied the anodization of tin in NaOH at constant potential [334]. They observed the formation of SnO pompon-like and flower-like microspheres comprised of SnO nanoflakes as a function of the potential, the anodization time and NaOH concentration (Figure 50). Superhydrophobic properties with $\theta_w = 153°$ were reported. Finally, Habazaki et al. reported the formation of niobium oxide microcones by anodization in hot phosphate (K_2HPO_4 and K_3PO_4)–glycerol electrolytes and at constant potential [335]. The size of the microcones and their tip angles could be controlled with the potential and the water content.

Figure 50 Pompon-like structures obtained by anodizing tin substrates in NaOH. Reprinted with permission from ref. 334, copyright 2012, Elsevier.

3.6 Electrodeposition

3.6.1 Noble Metals

Noble metals are not subjected to oxidation because their oxidation potential is much higher than that of H_2O/H_2 and O_2/H_2O couples. Hence, even if they are in contact with air or immersed in aqueous media, the formation of metal oxide or hydroxides is not favorable.

3.6.1.1 Silver

Ag dendritic structures were electrodeposited from $AgNO_3$ aqueous solution and at constant potential [336] (Figure 51). At −1 V, faceted nanostructured were deposited, while at −2 V, dendritic structures with high roughness were formed. After modification with a hydrophobic thiol, superhydrophobic properties were obtained ($\theta_w = 154.5°$ and $\alpha_w < 2°$). These materials were also found to have antibacterial properties [337]. Superhydrophobic silver nanoplates were also obtained using an aqueous solution of AgNO$_3$ and sodium citrate and at a current density of 5 mA cm^{-2} for 5 h [338]. However, here it was first necessary to deposit Ag nanoseeds before electrodeposition. Huang et al. electrodeposited Ag nanoparticles on colloidal polystyrene spheres in order to obtain micro- and nanostructures [339]. Here, it was first necessary to deposit a thin gold layer before electrodeposition to use the substrate as electrode. These substrates displayed superhydrophobic properties and were also used for SERS detection and could detect highly diluted molecules (femtomolar level).

Figure 51 Silver dendritic structures obtained by electrodeposition in AgNO$_3$. Reprinted with permission from ref. 336, copyright 2008, American Chemical Society.

3.6.1.2 Gold

Au nanostructures with various morphology, including dendritic rods, nanosheets, flower-like structures, and pinecone-like structures, were obtained by electrodepositing at constant current density in KAuCl$_4$ and H$_2$SO$_4$ [340–342]. The surface morphology was highly dependent on the current density and the deposition time. After modification with a hydrophobic thiol, properties were obtained (θ_w = 153.4° and α_w = 4.4°). Au flower-like structures were also reported by square wave voltammetry technique [341]. Here, the diameter and the density of the nanoflowers could be controlled by repetitive run times. Cai et al. reported regular arrays of nanoflowers by electrodeposition [343]. Here, colloidal polystyrene spheres were used as invisible template for the growth of nanospheres (Figure 52). The resulting substrates displayed superhydrophobic and could be used as sensors for

SERS applications. Jiang et al. reported also the possibility of producing Au microflowers on micro-pillar arrays along the solid–liquid–gas triphase interface [340].

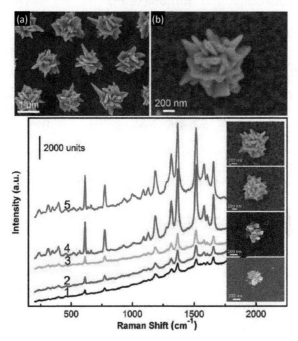

Figure 52 Array of gold nanostructures obtained by electrodeposition in HAuCl$_4$ and using colloidal polystyrene spheres as invisible template. The substrates were used for SERS applications. Reprinted with permission from ref. 343, copyright 2013, American Chemical Society.

3.6.1.3 Platinum and Palladium

Platinum and palladium electrodeposition was also studied in the literature. Kim et al. reported the formation of Pt tree-like structures composed of elongated trigonal pyramidal shapes with sharp edge sites on their side wall [344]. The authors used solutions of K$_2$PtCl$_4$ and H$_2$SO$_4$ and constant potential of −0.2 V to induce their growth. The density of the nanostructures could be controlled with the deposition charge. After post-treatment with a hydrophobic thiol, superhydrophobic properties (θ_w > 150°, α_w = 3.7° and H_w = 7.9°) with SERS activity were obtained. Similar results were obtained with Pd using K$_2$PdCl$_4$

[345] (Figure 53). Nanoflakes were electrodeposited using a constant potential of 0.20–0.25 V.

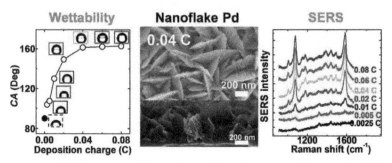

Figure 53 Palladium nanoflakes obtained by electrodeposition in K_2PdCl_4 and H_2SO_4. The substrates were used for SERS applications. Reprinted with permission from ref. 345, copyright 2015, American Chemical Society.

3.6.2 Non-Noble Metals

3.6.2.1 Copper

Non-noble metals can be oxidized to form metal oxides or hydroxides. In water, the general reaction is $M + H_2O \rightarrow M^+ + OH^- + H_2 \rightarrow M(OH)$ and or MO. The formation of metal oxides or hydroxides is highly dependent on the metal, more precisely on its oxidation potential, but also on the pH (metal are usually formed in acidic media while metal oxides and hydroxides in neutral and basic media). Metals are sensitive to water if the oxidation potential of the M^+/M couple is below that of H_2O/H_2 while the oxidation rapidity is dependent on the difference between the two potentials.

Among all non-noble metals, copper is one of the most used metals subjected to oxidation. Rough copper substrates were obtained by electrodeposition using $CuCl_2$ or $CuSO_4$ as copper source [346–349]. In acidic media, Cu is usually the main product while in basic media copper hydroxides $Cu(OH)_2$ and/or copper oxides CuO and Cu_2O can be obtained. The surface roughness can be controlled with many parameters such as the electrodeposition time. Then, superhydrophobic properties can be reached, for example, after post-treatment with a hydrophobic thiol or acid. These materials could be also electrodeposited on copper mesh substrates, leading to superhydrophobic and superoleophilic properties [350]. By electrodepositing rough Cu_2O, Li et al. reported high separation abilities (>90%) for different kind of oils. Moreover, the mesh substrates displayed also anticorrosion properties.

It is also possible to electrodeposit extremely well-defined Cu surface structures by controlling the electrodeposition process. Indeed, the pH solution or the electrodeposition method and parameters are fundamental to control surface structures with this method. For example, Liu et al. reported Cu zigzag microstrips by electrodeposition at the constant potential of 1.5 V in a solution containing only $CuCl_2$ [351] (Figure 54). The authors believe that their formation is due to Cu^{2+} electromigration to cathode, rapid nucleation, and slow growth of the nuclei. Moreover, Cu zigzag microstrips displayed strong adhesive strength to the Cu substrate. After modification with stearic acid, the substrates displayed superhydrophobic (θ_w = 162.5° and α_w = 2.8° after 30 min electrodeposition) and anticorrosion properties.

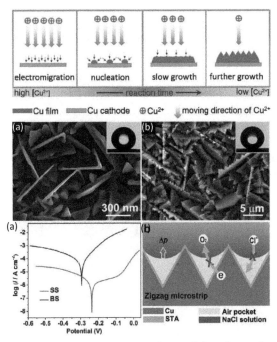

Figure 54 Copper zigzag microstrips obtained by electrodeposition in $CuCl_2$. The substrates were used in anticorrosion. Reprinted with permission from ref. 351, copyright 2015, Elsevier.

Cu cauliflower-like structures were reported by a two-step electrodeposition process: the first step at high overpotential (0.9 V) for the formation of structures with dense-branching morphology and the second step at low overpotential (0.15 V) for a short time to reinforce the loosely attached branches on the surface [352]. The substrates displayed superhydrophobic properties with low hysteresis (θ_w = 160° and H_w = 5°). The group of Li developed Cu nanocones arrays at a current density of 1.2 A dm^{-2} and using H_3BO_3 as pH buffer and Janus Green B as crystal modifier [353] (Figure 55). The growth of Cu nanocones is induced along the <111> direction and can be explained by the screw dislocation driven theory. The size of the nanocones could be controlled with the deposition temperature and time. The substrates were superhydrophilic, but after modification with a hydrophobic thiol they displayed superhydrophobic properties (θ_w = 164.3°). In order to produce micro- and nanostructures substrates, caterpillar-like structures were developed using a two-step electrodeposition process [354]. First, Cu nanocones array were prepared using a similar procedure. Then, nanofibers were induced on these nanocones using a second electrodeposition of Ni–Co alloy.

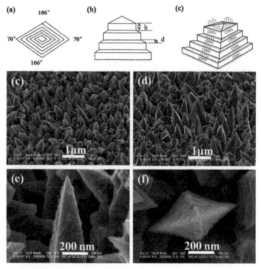

Figure 55 Copper nanocones obtained by electrodeposition in CuSO$_4$, H$_3$BO$_3$ as pH buffer and a crystal modifier. Reprinted with permission from ref. 353, copyright 2015, Royal Society of Chemistry.

In order to prepare microporous and nanostructured copper substrates, the dynamic hydrogen bubble templating process can be used. In this process, hydrogen bubbles arise from the electrochemical reduction of H$^+$ during the electrodeposition process and serve as dynamic template for metal electrodeposition. To produce H$_2$ bubbles, Li et al. used a solution of CuSO$_4$ and H$_2$SO$_4$ and a current density of 300 A dm^{-2} [355]. Superhydrophobic properties (θ_w = 162.1°) were obtained after modification with a hydrophobic thiol. They also showed that the pore diameter increases simultaneously with the deposition time but the contact angle decreases. The surface morphology could also be controlled using square wave potential pulses to form micro-coral structures [356].

The possibility of using dynamic hydrogen bubble templating process on copper mesh substrates was also reported [357–360]. It was possible to stabilize the hydrogen bubbles and control their size by adding a surfactant (CTAB) in the electrolyte, while the wall width was highly dependent on the applied current density and surfactant and CuSO$_4$ concentrations. The highest superhydrophobic properties (θ_w = 162°) were obtained with mesh pore sizes of 10 µm and without any surface hydrophobization. O'Mullane also used these substrates but modified them with 7,7,8,8-tetracyanoquinodimethane (TCNQ) and 2,3,5,6-tetrafluoro-7,7,8,8-tetracyanoquinodimethane (TCNQF$_4$) resulting in the formation of needle-like CuTCNQ and spike-like CuTCNF$_4$ copper complexes with superhydrophobic properties (θ_w up to 177°) [359] (Figure 56).

The surface chemistry can also be modified. For example, superhydrophobic Cu$_2$S rough surfaces were obtained by square-wave potential pulses in the presence of Na$_2$S$_2$O$_3$ [361]. Cu–Ni composites were also electrodeposited in the presence of CuSO$_4$, Ni(NH$_2$SO$_3$)$_2$, and H$_3$BO$_3$ [362]. The Cu and Ni ratios could be controlled with the CuSO$_4$ concentration and the applied voltage respectively leading to pillar- and dome-shaped structures with superhydrophobic properties. Cu–Ni composites were also prepared using the dynamic hydrogen bubble templating process using a current density of –70 to –400 A dm^{-2} in order to form microporous and nanostructured substrates [363]. The optimal Ni content was found to be 15 at%. The substrates displayed superhydrophobic properties, electrocatalytic activity, and ferromagnetic behavior with coercivity ranging from 114 to 300 Oe. These materials are excellent candidates for magnetically actuated micro/nano-electromechanical systems (MEMS/NEMS) or in magnetic sensors or separators. Cu–ZnO composites were also grown on Zn foils in the presence of $CuCl_2$, $ZnCl_2$, and H_2O_2 [364, 365]. A constant potential of −2.2 V vs. SCE was used. The substrates displayed superhydrophobic properties and different morphologies such as leaf-like, flower-like, and hexagonally funnel tubes were obtained.

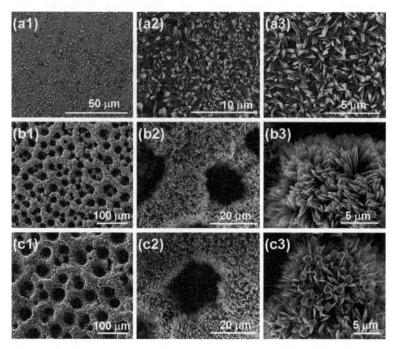

Figure 56 Microporous and nanostructured $TCNQF_4$ obtained with the dynamic hydrogen bubble templating (H_2 bubbles are produced in situ). Reprinted with permission from ref. 359, copyright 2013, Elsevier.

3.6.2.2 Others

Rough nickel substrates were prepared in solution containing $NiCl_2$ and H_3BO_3 [366, 367]. After post-treatment, it was possible to obtain superhydrophobic (θ_w = 160.8°, α_w = 1.8°) anticorrosion properties. The dynamic hydrogen bubble templating process was also used to formed microporous and nanostructured substrates. Here, the process was performed in the presence of $NiCl_2$, NH_4Cl and at pH 4 using an intensity of 3 A. After annealing, the substrate was used as a template and filled with PDMS to form a gecko-like dry adhesive [368].

In order to better control the surface structures, a deep eutectic solvent consisting of a mixture of choline chloride and ethylene glycol was used as the electrolyte [369]. Various structured architectures such as nanosheets, aligned nanostrips, and hierarchical flowers were obtained as a function of the electrodeposition method (constant potential, pulsed potential, and reverse pulse potential) [370, 371]. The highest superhydrophobic properties (θ_w = 164°, H_w = 4°) were reached with nanosheets obtained at a constant potential. Sponge-like structures made of interconnected nanofibers of nickel hydroxides $Ni(OH)_2$ were also produced in a basic solution using CH_3COONa. After post-treatment, the substrates displayed both high transparency and superhydrophobic properties with low water adhesion. It was also reported the possibility of reversibly switching from superhydrophobic to superhydrophilic by UV/ozone and heat treatments.

As reported for copper, several groups also reported the formation of Ni nanocones arrays at constant current using $NiCl_2$, H_3BO_3 as pH buffer and a crystal modifier or capping agent such as ethylenediammonium dichloride, NH_4Cl or NaCl [372–378]. The size and the morphology of the nanocones can be controlled with the current density, the deposition time, or the crystal modifier. It was also possible to enhance the superhydrophobic properties by using a two-step electrodeposition process: the first at 20 mA cm^{-2} for 600 s to form microstructures and the second at 50 mA cm^{-2} for 60 s to form nanocones. The substrates also displayed anticorrosion properties. Substrates with nanocone arrays were used as template and filled with gold [378] (Figure 57). After removal of the template in HNO_3, the resulting substrates had inverted conical holes with a diameter of about 300 nm and a depth of about 800 nm. Moreover, they displayed superhydrophobic properties and were used for SERS detection. Micro-nano flower-like structures of Ni–Co alloys were also reported using a similar procedure but by adding $CoCl_2$ in the electrolyte [379–382].

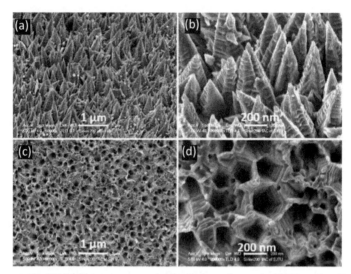

Figure 57 Nickel nanocones obtained by electrodeposition in NiCl$_2$, H$_3$BO$_3$ as pH buffer and a crystal modifier. The substrates were used to transfer nanostructures. Reprinted with permission from ref. 378, copyright 2015, American Chemical Society.

The electrodeposition of NiCl$_3$ in the presence of dodecanoic acid (CH$_3$(CH$_2$)$_{10}$COOH) in ethanol led to with micro/nano papillae structures of nickel dodecanoate (Ni(CH$_3$(CH$_2$)$_{10}$COOH)$_3$ [383] (Figure 58). The equations are

$$Ni^{3+} + 3\,CH_3(CH_2)_{10}COOH \rightarrow$$
$$Ni(CH_3(CH_2)_{10}COOH)_3 + 3\,H^+\; 3\,H^+ + 2\,e^- \rightarrow H_2.$$

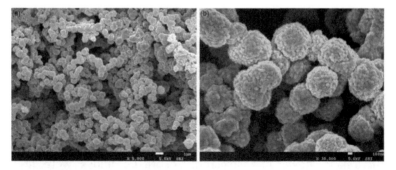

Figure 58 Fe(CH$_3$(CH$_2$)$_{14}$COOH)$_3$ nanostructured microspheres obtained by electrodeposition in FeCl$_3$ and CH$_3$(CH$_2$)$_{14}$COOH. Reprinted with permission from ref. 384, copyright 2016, Elsevier.

Superhydrophobic properties were obtained with this process. Similar results were obtained with iron and manganese using FeCl$_3$ and MnCl$_2$, respectively [384–386].

The formation of different nanostructures of cobalt by electrodeposition was also reported [387–390]. For example, using CoCl$_2$ and Na$_2$SO$_4$, superhydrophobic flower-like structures were observed at −1 V vs. SCE while superhydrophilic dendritic structures became predominant at −1.2 V [390]. A similar procedure was also used to electrodeposit tree-like structures on self-assembled gold nanoparticle–modified electrodes.

Co nanocones and nanoshells arrays were also reported at constant current using CoCl$_2$, H$_3$BO$_3$ as pH buffer, and a crystal modifier [391] (Figure 59). The nanocones and nanoshells were fabricated with 10 A dm^{-2} for 1 min and 1.25 A dm^{-2} for 20 min, respectively. The nanoshells displayed superhydrophobic properties θ_w = 156° and α_w = 10°, while the nanocones displayed parahydrophobic properties with θ_w = 154° but extremely high water adhesion.

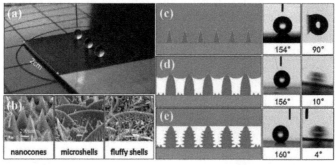

Figure 59 Cobalt naocones, microsheels, and fluffy shells obtained by electrodeposition in CoCl$_2$, H$_3$BO$_3$ as pH buffer and a crystal modifier and using different current densities. Reprinted with permission from ref. 391, copyright 2015, Elsevier.

Superhydrophobic structured ZnO substrates with various morphologies were prepared by electrodeposition using a zinc salt and using different deposition methods, and heat treatment [392–394]. ZnO was also electrodeposited in the interstice of colloidal polystyrene spheres used as removal template [395]. After dissolution, a ZnO inverse opal structure was obtained displaying superhydrophobic properties and suitable for electrowetting experiments.

Other research groups were highly interested in the wettability of ZnO nanorod arrays obtained by electrodeposition. Pauporté et al. reported the preparation of ZnO nanorod arrays at constant potential and using aqueous solutions of $ZnCl_2$ and KCl, saturated with O_2 by bubbling [396–399]. Moreover, the aspect ratio of the nanorods could be easily controlled, for example, with the deposition temperature, the deposition time or $ZnCl_2$ concentration. After surface treatment, the authors found that the length of the nanorods is a key factor in the formation of superhydrophobic properties. Indeed, as the ZnO nanorod length increased, a water droplet switched from Wenzel to Cassie–Baxter state (Figure 13c). Moreover, the density of nanorods was also an important parameter in the control of superhydrophobic properties. The control of the surface wettability was also possible using different acids as post-treatment but also by redox changes of ferrocene molecules grafted on the substrate [400, 401]. The alignment and the control of the density could also be performed by depositing ZnO seed layer before electrodeposition.

Rough ZrO_2 and WO_3 were also reported by using $ZrClO_2$ and Na_2WO_4 as metal source [402–404]. Sn/SnO_x dendritic structures were prepared using the dynamic hydrogen bubble templating process [405] (Figure 60). A solution of $SnCl_2$ and H_2SO_4 was used and the electrodeposition was performed at constant potential of –1.6 to 1.9 V. The presence of both microporosity and dendritic structures led to superhydrophobic properties (θ_w = 165°) with surface post-treatment. Moreover, the surface properties could be controlled with $ZnCl_2$ concentration and the applied potential.

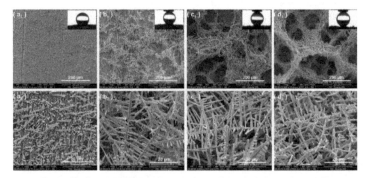

Figure 60 Sn/SnO_x dendritic structures obtained by electrodeposition in $SnCl_2$ and H_2SO_4 and using the dynamic hydrogen bubble templating process. Reprinted with permission from ref. 405, copyright 2013, Elsevier.

In order to improve the oxidation capability of Pb_2O electrodes, superhydrophobic Pb_2O substrates were obtained by electrodeposition on PTFE doped Pb_2O on TiO_2 nanotubes using solution of $Pb(NO_3)_2$, HNO_3, KF, and PTFE as electrolytes [406, 407].

3.6.3 Rare Earths

Rare earth superhydrophobic coatings were also reported by electrodeposition. Rough substrates made of spherical particles were prepared in the presence of $Ce(NO_3)_3$ and myristic acid $(CH_3(CH_2)_{12}COOH)$ in ethanol [408–416]. The experiments were performed at a constant potential of 30 V and the surface roughness could be controlled with the deposition time and the applied voltage. In this process, the Ce^{3+} ions near the cathode reacted with myristic acid to form cerium myristate $Ce(CH_3(CH_2)_{12}COOH)_3$ complexes (Ce^{3+} + 3 $CH_3(CH_2)_{12}COOH \rightarrow Ce(CH_3(CH_2)_{12}COOH)_3$ + 3 H^+). The surfaces were superhydrophobic (θ_w = 159.8° and α_w < 2°) and also resistance to corrosion. $CeCl_3$ was also used as cerium source. For example, micro-nano flower-like structures displaying superhydrophobic properties (θ_w = 162° and α_w < 4°) were reported. Similar results were also obtained with lanthanum [417] (Figure 61).

Figure 61 $La(CH_3(CH_2)_{12}COOH)_3$ flower-like structures obtained by electrodeposition in $LaCl_3$ and $CH_3(CH_2)_{12}COOH$. Reprinted with permission from ref. 417, copyright 2012, Elsevier.

3.6.4 Sol-Gel Electrodeposition

SiO_2 and TiO_2 nanoparticles are usually prepared using the sol-gel process and can be applied on substrates by dip-coating or spin-coating. Very recently, Hu et al. showed that they can be also deposited by electrodeposition using, for example, tetraethoxysilane (TEOS) as silicon source (Figure 62) [418–422]. Here, the key factor is the formation of OH^- which act as catalyst for the reaction and generate SiO_2 precipitation. When a negative potential is applied, the reactions are, for example,

$$2\, H_2O + 2\, e^- \rightarrow H_2 + 2\, OH^-$$
$$NO_3^- + H_2O + 2\, e^- \rightarrow NO_2^- + 2\, OH^-.$$

The electrodeposited films are thicker and rougher than those obtained by the conventional sol-gel process and with greater porosity. Superhydrophobic properties were reported with also anticorrosion properties.

Figure 62 SiO_2 nanoparticles obtained by electrodeposition in tetraethylorthosilicate (TEOS). Reprinted with permission from ref. 420, copyright 2016, Elsevier.

3.7 Electroless Deposition

The electroless deposition or galvanic replacement is a spontaneous reaction between a metal substrate immersed into an aqueous solution containing ion metals of another metal. The reaction is $M_1 + M_2^+ \rightarrow M_1^+ + M_2$. In this reaction, the metal M_1 is oxidized, whereas M_2^+ ions are reduced. This reaction is possible only if the oxidation potential $E^O_{M^{2+}/M_2}$ is higher than $E^O_{M^{1+}/M_1}$. Moreover, this reaction is spontaneous only if the difference between the two potentials, $\Delta E = E^O_{M^{2+}/M_2} - E^O_{M^{1+}/M_1}$, is high (typically above 0.5 V).

3.7.1 Silver

The most famous example is the reaction between Ag$^+$ (for example, AgNO$_3$) and copper substrates. It often led to the formation of dendritic Ag fractal structures [423–434]. The surface roughness could be controlled with AgNO$_3$ concentration and the immersion time (Figure 63). Superhydrophobic properties could be easily obtained after post-treatment and even superoleophobic properties using fluorinated compounds. In order to enhance the nonwettable properties, the copper substrate can be first structured, for example, by etching or oxidation before electroless deposition [435, 436]. He et al. first deposited CuO nanoneedles on microplotted substrates and then Ag dendrites were grown on these substrates [345]. Albadarin et al. also developed copper meshes with dendritic Ag structures [437]. Then, they used mercaptoundecanoic acid (HS(CH$_2$)$_{10}$COOH) in order to control the water permeation and organic solvents/water separation with a high separation efficiency (99.8%) by adjusting the pH.

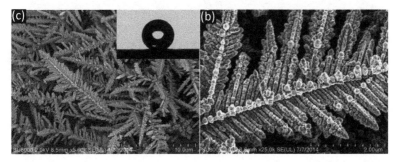

Figure 63 Silver dendritic structures obtained by immersing copper substrate in AgNO$_3$. Reprinted with permission from ref. 431, copyright 2015, Wiley.

Different compounds can also be added in solution in order to change the surface morphology [438–442]. For example, Yao et al. reported the formation of Ag flower-like microstructures composed of nanoplates passing through [Ag(NH$_3$)$_2$]OH complex obtained between Ag$^+$ and ammonia [438]. Muench et al. added different amount of Cl$^-$ and Br$^-$ ions in the solution [441] (halide-assisted electroless deposition) (Figure 64). However, in order to avoid the precipitation of AgCl and AgBr, ethylenediamine (H$_2$N-CH$_2$-NH$_2$) was used as complexing agent. With this process, particles of various shapes, for example, with triangular or hexagonal shape were observed leading to superhydrophobic properties (θ_w = 165° and α_w < 3°). Microballs composed of nanoplates were also reported in the presence of ammonium citrate and NaH$_2$PO$_2$ as a reducing agent [442] (Figure 65).

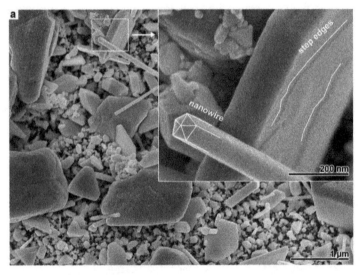

Figure 64 Silver flower-like microstructures composed of nanoplates obtained using the halide-assisted electroless deposition. Reprinted with permission from ref. 441, copyright 2015, Royal Society of Chemistry.

Other metals can be used for electroless deposition with Ag$^+$ [443–446]. Dendritic structures were also reported on stainless steel. Dendritic structures and horn-like rods were deposited on zinc substrates by immersion in AgNO$_3$ [444]. The authors reported the influence of AgNO$_3$ concentration, deposition time, and the pretreating temperatures on the surface morphology. Superoleophobic properties were obtained after modification with a perfluorinated thiol ($\theta_{hexadecane}$ = 138°). Silver nanoplates were also reported by immersion of GaAs substrates in AgNO$_3$.

The reaction proceeds by oxidation of gallium and arsenic to form Ga$_2$O$_3$ and As$_2$O$_3$ while Ag$^+$ is reduced in Ag [445, 446]. Superhydrophobic surfaces with low sliding angles were obtained for a surface coverage with the nanoplates of above ≈30%.

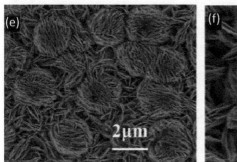

Figure 65 Nanostructured microballs by immersing copper substrate in AgNO$_3$, ammonium citrate and NaH$_2$PO$_2$ as reducing agent. Reprinted with permission from ref. 442, copyright 2013, Royal Society of Chemistry.

Aluminum substrates can also be used but it is necessary to added F$^-$ ions in the solution to first dissolve Al$_2$O$_3$ barrier oxide layer:

$$Al_2O_3 + 6\,H^+ + 12\,F^- \rightarrow AlF_6^{3-} + 3\,H_2O$$

Using this process and by adding NaF, Ag dendritic structures were obtained [447]. This process was also applicable to other metals with oxide layers such as iron, cobalt, and molybdenum. Wong et al. also used this process with HF to deposit Ag nanostructures on Al micropowders [448]. Different structures could be controlled with Ag$^+$ concentration. The resulting substrates were superhydrophobic and could be used for SERS.

Silicon substrates can also react using F$^-$ ions [449–450]. In this process, the silicon etching and Ag deposition occur simultaneously:

$$Si + 6\,HF + AgNO_3 \rightarrow H_2SiF_6 + Ag + 2H_2O + NO$$

Silver nanoclusters and dendritic structures can be controlled with the deposition time and AgNO$_3$ concentration. Superhydrophobic properties were reported with this process ($\theta_w = 152°$ and $\alpha_w \approx 8°$).

3.7.2 Gold and Platinum

Many other metals can be deposited with this method. Gold dendritic structures were obtained by electroless deposition of zinc substrate [451–453]. After etching in NaOH, the reaction was performed by immersing the substrate in an aqueous solution of HAuCl$_4$ for a certain time (Figure 66). After annealing at 160°, the presence of Au-Zn alloys (including AuZn$_3$ and AuZn) and ZnO were determined [451]. The resulting substrates were superhydrophobic (θ_w = 170° and α_w < 1°) without post-treatment. Gold dendritic substrates were also observed on silicon but using a mixture of HF and HAuCl$_4$ [452]. Cho et al. developed on these substrates switchable surface after modification with chloride groups and the polymerization of the surface by atom transfer radical polymerization (ATRP) to introduce a polymer with ammonium groups. Here, the substrates wettability could change from superhydrophobic to superhydrophilic after ions exchange.

Hierarchical platinum structures were prepared by immersing copper substrates in an aqueous solution of PtCl$_4$ and annealing [454]. The authors observed the presence of CuO and Cu$_3$Pt with superhydrophobic properties (θ_w = 170° and α_w ≈ 0°). Flower-like micro- and nanostructures were also reported by immersion of zinc substrates in PtCl$_4$ [455].

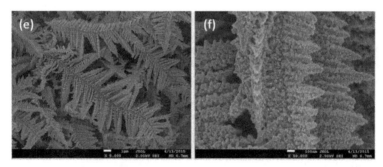

Figure 66 Gold dendritic structures obtained by immersing zinc substrate in HAuCl$_4$. Reprinted with permission from ref. 451, copyright 2015, Royal Society of Chemistry.

3.7.3 Others

Copper with various morphologies was also prepared by electroless deposition. For example, flower-like structures were reported on zinc substrates by immersion in CuSO$_4$ [456]. After annealing and post-treatment superhydrophobic properties were obtained (θ_w = 152° and α_w < 5°). Cu dendritic structures were also prepared by immersion in $CuCl_2$ and Na_2SO_4. Zhang et al. induced the growth of carbon fibers by CVD and obtained substrates with superhydrophobicity and anticorrosion properties [457]. Cu dendrites were also reported on aluminum substrate after etching in NaOH and immersion in $CuCl_2$ [458, 459]. Using a similar procedure, CuSe lamellar structures were obtained by immersing steel substrates in $CuSO_4$ and H_2SeO_3 solutions [460].

Zhang et al. reported density packed particles of tin by etching zinc substrates in HCl, immersing in $SnCl_2$ and annealing [461]. CeO_2 nanoparticles were also obtained by immersion in $CeCl_3$ and H_2O_2 [462]. Using a similar strategy, Bi_2O_3 dendritic structures were obtained using $BiCl_3$ and HCl [463, 464] (Figure 67). Finally, Li et al. developed a method to fabricate Co_3Ni leaf-like microstructures by immersion of copper substrates in the presence of $CoCl_2$, $NiCl_2$, triethylenetetramine, and sodium dodecylsulfate [465] (Figure 68). Then, a mixture of NaOH, hydrazine hydrate, and ethanol was added to the mixture to induce the formation of these leaf-like structures. The substrates displayed superhydrophobic properties, ferromagnetic properties with enhanced coercivity, and high catalytic performance in the hydrolysis of ammonia borane.

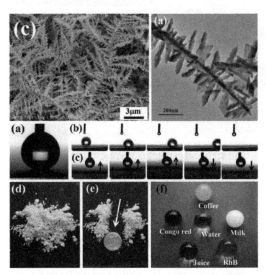

Figure 67 Bi_2O_3 dendritic structures obtained by immersing zinc substrate in $BiCl_3$ and HCl. The substrates displayed switchable wettability. Reprinted with permission from ref. 463, copyright 2015, Elsevier.

Figure 68 Co₃Ni leaf-like structures obtained by immersing copper substrate in CoCl₂, NiCl₂, a crystal modifier and sodium dodecylsulfate. Reprinted with permission from ref. 465, copyright 2015, Royal Society of Chemistry.

3.8 Hydrothermal Processes

A hydrothermal process is the crystallization of chemical compounds usually by reaction at high temperature.

3.8.1 Zinc Oxide

Among all crystalline structures developed by hydrothermal process, ZnO is probably the most studied material, for example, for the wide range of crystalline structure such as nanoplatelets, nanowires, and nanorods. In particular, highly dense vertically aligned nanorods have attracted many research groups for their exceptional wettability properties. The parameter of the ZnO nanorods can be easily controlled, for example, with the process time while the surface morphology can be controlled with the compounds concentrations [466–474] (Figure 69). For example, ZnO nanorods can be obtained using zinc nitrate $Zn(NO_3)_2$ and hexamethylenetetramine ($C_6H_{12}N_4$) following these reactions:

$$C_6H_{12}N_4 + 6\,H_2O \rightarrow 4\,NH_3 + 6\,HCHO$$

$$NH_3 + H_2O \rightarrow NH_4^+ + OH^-$$

$$Zn(NO_3)_2 \rightarrow Zn^{2+} + 2\,NO_3^-$$

$$Zn^{2+} + OH^- \rightarrow Zn(OH)_2 \rightarrow ZnO + H_2O$$

Figure 69 ZnO nanowire arrays grown from $Zn(NO_3)_2$ and hexamethylenetetramine. Reprinted with permission from ref. 469, copyright 2013, Wiley.

The most employed method to develop highly dense vertically aligned ZnO nanorods is the preparation of a ZnO nanoseed layer before the hydrothermal process. For example, Rajebdrakumar et al. studied the formation of ZnO nanoseed layers by spray pyrolysis and observed that the surface morphology is highly dependent on the temperature of the seed formation [466]. More precisely, vertically aligned nanorods were obtained at 550°C but the substrates were highly hydrophobic (θ_w = 119°) while superhydrophobic interconnected nanorods (θ_w = 152°) were obtained at 350°C (Figure 70).

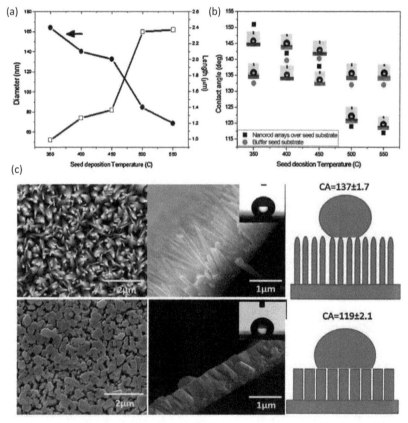

Figure 70 ZnO interconnected nanorods and vertically aligned nanorods grown from $Zn(NO_3)_2$ and hexamethylenetetramine, as a function of the temperature synthesis. Reprinted with permission from ref. 466, copyright 2014, Elsevier.

Li et al. compared this method with the formation of seed layer by atomic layer deposition (ALD) [467] (Figure 71). By pyrolysis the preferential growth is in the (002) crystallographic plane and nanorods are formed, while by ALD due to the presence of ethyl or methyl groups, the preferential growth is in the (001) crystallographic plane and superhydrophobic flower-like structures ($\theta_w = 170°$) are obtained. Lei et al. reported similar results but after replacing the zinc source by $Zn(NO_3)_2$ or $Zn(CH_3COO)_2$ [472]. Indeed, other authors reported the importance of the zinc source on the surface morphology.

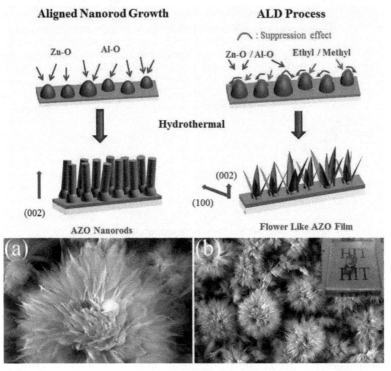

Figure 71 Comparison between hydrothermal process and atomic layer deposition [467].

Hence, ZnO nanorods arrays are highly hydrophobic with high water adhesion and it is possible to obtain superhydrophobic properties with low adhesion just by a post-treatment with a fluorinated silane, or an alkanoic acid or a hydrophobic polymer (PTFE, PDMS, etc.) [475–480]. However, Gao et al. showed that contrary to nanowires, the substrates with vertically aligned nanorods lost their low adhesion capacity after water droplet impacts [476]. By contrast, Ren et al. showed that stable superhydrophobic properties with low adhesion and anticorrosion properties can be obtained by coating with PDMS layer [479] (Figure 72).

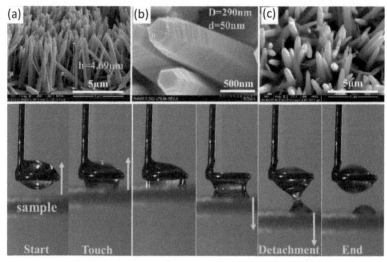

Figure 72 Superhydrophobic properties with low adhesion for ZnO nanorod arrays coated by PDMS. Reprinted with permission from ref. 479, copyright 2016, Royal Society of Chemistry.

Jiang et al. also used a second hydrothermal process to deposit CeO_2 [481]. They observed a complete change in the surface morphology from nanopencil arrays into nanotube arrays (Figure 73). The resulting substrates displayed superhydrophobic properties with θ_w = 157.3° and an adhesive force of 14 µN.

Otherwise, in order to obtain robust superhydrophobic properties, it is preferable to obtain both micro- and nanostructures. This could be achieved by adding different nanostructured materials or surfactants in order to induce multiscale hierarchical nanostructures [482–484]. Different works reported also the growth of ZnO nanostructures on microstructured substrates [485–499] (Figure 74). For example, natural superhydrophobic leaves were used as a microstructured pattern. Cortese et al. developed micropatterned PDMS substrates by photolithography and polymer casting [487]. After the growth of ZnO nanorod arrays, the substrates displayed superhydrophobic and underwater superoleophobic properties, indicating it as an excellent candidate for oil/water separation membranes. ZnO nanorod arrays were also studied on polystyrene spheres obtained by colloidal lithography [494]. The substrates displayed superhydrophobic properties with θ_w = 152°. Similarly, ZnO nanowires deposited on stainless steel mesh substrates of about 30 μm of wire thickness and gap displayed superhydrophobic properties with θ_w = 160.4° [495]. Other research groups also studied their growth on TiO_2 nanotubes obtained by titanium anodization and zinc hexagonal microstructures obtained by electrodeposition [496]. After fluorination, these last substrates displayed superoleophobic properties.

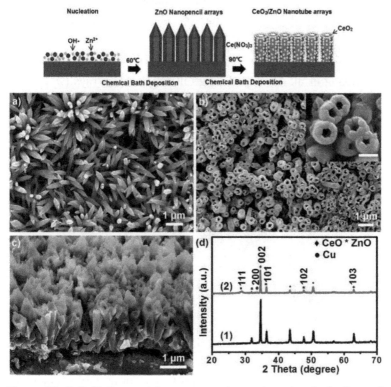

Figure 73 CeO_2/ZnO nanotube arrays using a two-step hydrothermal process. Reprinted with permission from ref. 481, copyright 2016, Elsevier.

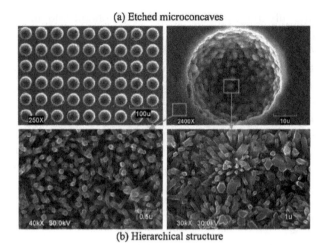

(a) Etched microconcaves

(b) Hierarchical structure

Figure 74 Superhydrophobic surfaces obtained by growth of ZnO nanorods on etched microconcaves. Reprinted with permission from ref. 488, copyright 2015, Elsevier.

Huang et al. developed long CuO nanowire arrays (Ø ≈ 100–200 nm and length ≈ 10 µm) by heating Cu foils at 500°C [498] (Figure 75). After hydrothermal growth of ZnO nanorods on

Figure 75 Hierarchical nanotree arrays by growth of ZnO nanorods on CuO long nanowires. The substrates displayed self-cleaning properties. Reprinted with permission from ref. 498, copyright 2011, American Chemical Society.

these CuO long nanowires, hierarchical nanotrees array were obtained. After modification with a perfluorinated silane, the substrates displayed superhydrophobic properties ($\theta_w \approx 170°$) with ultra-low water adhesion (high resistance to water droplet impacts) and self-cleaning properties.

3.8.2 Others

Many others nanostructured oxides and hydroxides and even other materials were used to obtain superhydrophobic properties by hydrothermal process [500–526]. For example, copper [500–503], magnesium [504], aluminum [505–508], iron [509, 510], nickel [511, 512], cobalt [513], titanium [514–521], and lanthanum [522] were used in the literature, but also bimetallic oxides or hydroxides such as $CaTiO_3$ [523–526].

Nanostructured Eu^{3+}-doped $NaLa(MoO_4)_2$ was prepared by mixing Na_2MoO_4, $LaCl_3$, and $EuCl_3$ with a complexion and capping agent [527]. The resulting substrates displayed both superhydrophobic ($\theta_w = 162°$) and fluorescence properties. Sulfides and selenides were also reported using sulfur or selenium source. For example, superhydrophobic ($\theta_w > 160°$) $Cu_{2-x}Se$ nanostructures such as nanosheets and nanoribbons were produced by immersing Cu substrates in solution containing Se powder, NaOH, ethylenediamine or ammoniac [528] (Figure 76). The surface nanostructures could be controlled find several parameters such as reaction temperature, time or the concentration of NaOH. Quantum dots such as CdS were also reported using a solvothermal process in ethanol and using cadmium acetate and sodium sulfide [529]. After modification with dodecanethiol, the substrates displayed superhydrophobic and fluorescence properties. The surface properties could be controlled with the Cd/S molar ratio, the reaction temperature and time. Finally, nitrides such as BN were also reported by hydrothermal process and exfoliation. [530, 531] Highly hydrophobic ($\theta_w = 140°$) BN nanoflowers composed of vertically aligned nanoflakes were obtained following the reaction at 300°C:

$NaBF_4 + 3 NH_4Cl + 3 NaN_3$
$\rightarrow BN + NaF + 3 NaCl + 4 N_2 + 3 NH_3 + 3 HF$

76 | Fabrication Processes

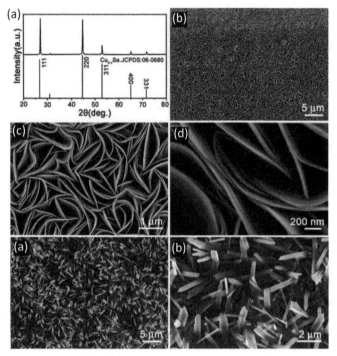

Figure 76 $Cu_{2-x}Se$ nanosheets and nanoribbons obtained by immersing copper substrates in solution containing Se powder, NaOH, ethylenediamine or ammoniac. Reprinted with permission from ref. 528, copyright 2011, Royal Society of Chemistry.

3.8.3 Applications

3.8.3.1 Photoluminescence

Moreover, the applications of superhydrophobic ZnO are numerous. ZnO, being a semiconductor, one of its properties is light absorption, for example, UV irradiation. Their absorption induces the formation of hydroxyl groups on the surface and as a consequence it can lead to superhydrophilic properties. The absorption of UV irradiation can induce photoluminescence [532, 533]. Ozer studied the growth of ZnO nanorods and nanowires on cotton fibers [534]. The authors observed higher photoluminescence properties in the deep UV region for the nanowires indicating fewer nonradiative transitions between excitonic states. Moreover, a blueshift in the emission and absorption onset was also observed for the nanoneedles, which was attributed to a decrease in structural defects and a widened band gap. Other materials also can display photoluminescence properties. Similarly, superhydrophobic ZnS nanowires and nanorods were also prepared by hydrothermal process in the presence of thiourea [535]. The higher photoluminescence properties were observed for ZnS nanowires, which also displayed strong yellow emission peaks.

3.8.3.2 Photocatalytic properties

Another application induced by the UV absorption is the photocatalytic activity. The photocatalytic activity allows the destruction of organic molecules and can be used for water cleaning [536, 537]. Bhattacharya et al. showed that ZnO nanorods arrays displayed also photocatalytic properties. When exposed to UV light, these substrates were able to degrade rhodamine dye in about 400 min. The authors believe that the high photocatalytic activity results from the high state of defect density and also high surface area. In(OH)$_3$ nanocubes were obtained by hydrothermal process in the presence of In(NO$_3$)$_3$ and proline and NaOH [538]. After fluorination, the substrates displayed superhydrophobic properties (θ_w = 152.2° and α_w = 1°) and photocatalytic activity against rhodamine B under UV illumination (93.25% degradation after 16 h). Bi$_2$WO$_6$ nanosheets prepared from Bi(NO$_3$)$_3$ and Na$_2$WO$_4$ and at pH 9 were also studied [539, 540]. The substrates displayed superhydrophobic properties (θ_w = 160.7°) and photocatalytic activity against methylene blue under visible illumination (93% degradation after 3 h). Similar results were obtained with Sb$_2$WO$_6$ [541] (Figure 77).

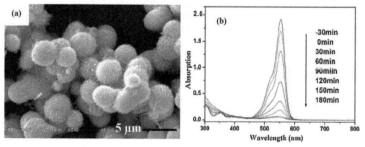

Figure 77 Sb$_2$WO$_6$ microspheres obtained from SbCl$_3$, Na$_2$WO$_4$ and NaOH. The substrates displayed photocatalytic properties. Reprinted with permission from ref. 541, copyright 2016, Elsevier.

Other authors also reported the superhydrophobic and photocatalytic properties of CeVO$_4$ nanoplates [542] and zinc phtalocyanine hollow microrectangular tubes [543]. Different titanium phosphates such as Ti$_2$O$_3$(H$_3$PO$_4$)$_2 \cdot$2H$_2$O, α-titanium Ti(HPO$_4$)$_2 \cdot$H$_2$O, and π-titanium Ti$_2$O(PO$_4$)$_2 \cdot$H$_2$O with various morphologies (nanobelts, microflowers, nanosheets, nanorods, nanoplates) were prepared by immersing titanium in aqueous solutions of H$_2$O$_2$ and H$_3$PO$_4$ at high temperature [544] (Figure 78). After post-treatment using alkylamine, water adhesion can be controlled with the alkyl chain length. Moreover, under UV illumination, their photocatalytic properties induced the degradation of the alkylamine and as a consequence a modification of the water adhesion.

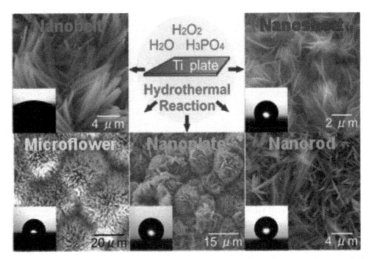

Figure 78 Various nanostructured titanium phosphates obtained by immersing titanium in aqueous solutions of H$_2$O$_2$ and H$_3$PO$_4$ at high temperature. Reprinted with permission from ref. 544, copyright 2014, American Chemical Society.

3.8.3.3 Others

These kinds of materials have also other properties extremely interesting for various potential applications. For example, it was shown that ZnO nanostructures such as nanorods or nanowires have antireflection properties and can be used in solar cells [545] (Figure 79). Hiralal et al. showed that ZnO nanowires can increase the efficiency of solar cells by 36% [546].

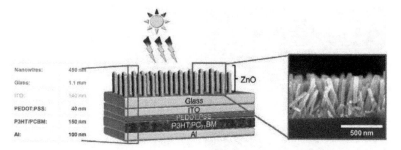

Figure 79 ZnO nanorod arrays using in solar cells. Reprinted with permission from ref. 545, copyright 2016, Elsevier.

Other authors were also interested in their electrowetting properties. Wang et al. studied the electrowetting of ZnO nanowires coated by a PTFE layer [547]. When the applied voltage was less than 50 V θ_w changed reversely from 165° to 120°, while at a higher voltage the change was more important but not reversible. The group of Tu also studied WO_3 nanowires and showed a decrease of θ_w from 159° to 140 for a voltage of 15 V [548]. These materials can also be used for application in sensors. Lee et al. demonstrated the superhydrophobic and H_2S gas-sensing properties of nanostructured CuO, following this reaction:

$$CuO + H_2S \rightarrow CuS + H_2O \text{ [549]}$$

A response of ≈34% was determined with response and recovery times to 2 ppm H_2S of 107 and 127 s, respectively. After inserting Ag nanoparticles, ZnO nanostructured substrates were also used for SERS [550] (Figure 80). Using R6G, the authors observed an enhancement factor value of about 3 × 10^6.

The group of He reported the catalytic effect of Co_3O_4 containing carbon nanotubes [551]. Here, the authors used cobalt (II) acetate and ammonium hydroxide to produce Co_3O_4 spherical nanoparticles. With the presence of carbon nanotubes and after fluorination, the substrates displayed highly hydrophobic properties (θ_w = 142.4°) and high catalytic activity for CO oxidation (Figure 81). Moreover, the authors observed the increase in the surface hydrophobicity of carbon nanotubes allows the enhancement of the moisture resistance of metal oxide catalysts for CO oxidation.

80 | Fabrication Processes

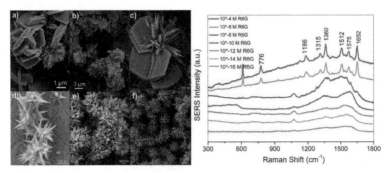

Figure 80 ZnO/Ag urchin-like structures used for surface-enhanced Raman detection (SERS). Reprinted with permission from ref. 550, copyright 2013, Royal Society of Chemistry.

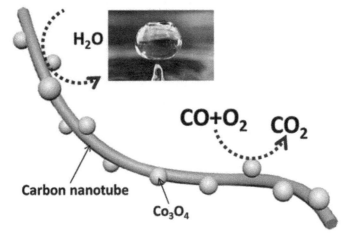

Figure 81 Carbon nanotubes/Co_3O_4 with high catalytic activity for CO oxidation. Reprinted with permission from ref. 551, copyright 2014, American Chemical Society.

Several works also reported the anticorrosion properties of oxides or hydroxides obtained after the hydrothermal process and surface modification. For example, superhydrophobic ZnO nanoplates were found to reduce about 23-fold from the bare aluminum substrate [552, 553]. Indeed, ZnO nanoplates showed a lower corrosion potential and a lower current density. This is probably due to the presence of air trapped inside the nanostructures, which prevents electron transfer between the electrolyte and the substrate. Moreover, ZnO is a semiconductor material and as a consequence has much higher resistance than aluminum. Cerium oxide nanorods were also studied [554, 555]. The substrates displayed both superhydrophobic properties (θ_w = 160.4° and α_w = 5.0) and anticorrosion properties. Magnesium being a metal highly subject to corrosion was also studied in the literature. It was shown that the formation of Zn-Al or Mg-Al double hydroxide nanosheets as well as szaibelyite ($Mg(BO_2(OH))$) fibrous structures can lead to superhydrophobic properties and also reduce the magnesium corrosion [556–558]. $MnWO_4$:Dy^{3+} microbouquets and Cu_9S_5 microprotrusions were also found to lead to both superhydrophobic and anticorrosion properties [559, 560] (Figure 82).

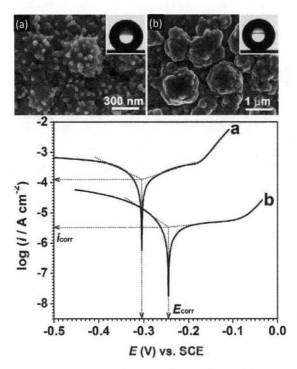

Figure 82 Cu_9S_5 nanostructured microspheres obtained by immersion of copper substrates in $CuCl_2$ and $Na_2S_2O_3$. The substrates displayed anticorrosion properties. Reprinted with permission from ref. 560, copyright 2015, Elsevier.

Otherwise, these materials can also be used for oil/water separation or for oil sorbent [561–563].

3.9 Use of Nanoparticles

3.9.1 Nanoparticles

The use of nanoparticles is one of the easiest ways to produce transparent superhydrophobic and even sometimes superoleophobic properties with high resistance against various environments. The most employed nanoparticles are SiO_2 nanoparticles. SiO_2 nanoparticles of different sizes can be produced using the Stöber reaction, also called the sol-gel process, in the presence of a silicon source such as tetraethoxyorthosilicate (TEOS) (Figure 83). Then, they can be applied on substrates by dip-coating, spin-coating and spraying, for example [564–579].

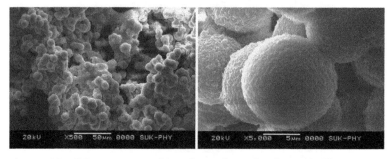

Figure 83 SiO_2 nanoparticles obtained with the sol-gel process. Reprinted with permission from ref. 570, copyright 2011, Elsevier.

For example, Motlagh et al. obtained superoleophobic properties by mixing SiO_2 nanoparticles of different sizes, following by a post-treatment with a fluorinated silane [546, 565]. The multiscale roughness induced by mixing nanoparticles of different sizes highly increased the oil apparent contact angle and decreased the oil sliding angle. The multiscale roughness was also confirmed by He et al., who mixed different kinds of silica particles, including 20 nm silica nanoparticles, 60 nm hollow silica nanoparticles, and silica nanosheets, to obtain superoleophobic properties ($\theta_{hexadecane}$ = 132.4°) but with high oil adhesion [572–573]. In order to produce superhydrophobic or superoleophobic properties in one-step, fluorinated SiO_2 nanoparticles can also be produced [579]. Moreover, the stability and mechanical properties can also be enhanced by grafting nanoparticles each other or to substrates using coupling agents.

The surface morphology can also be controlled with many parameters. He et al. reported an etching process with NaBH$_4$ in order to induce nanoroughness on the surface of SiO$_2$ nanoparticles [578] (Figure 84). Using nanoparticles of 240 nm, the authors reported the obtaining of superhydrophobic properties with $\theta_w = 150°$, $H_w = 3°$ and $\alpha_w < 1°$.

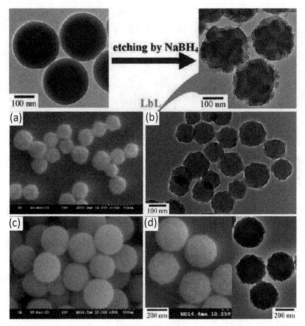

Figure 84 Formation of nanoroughness on SiO$_2$ nanoparticles by chemical etching. Reprinted with permission from ref. 578, copyright 2011, American Chemical Society.

Various other nanoparticles can be employed to induce sufficient roughness and lead to superhydrophobic properties. For example, TiO$_2$ [580–583], Al$_2$O$_3$ [584, 585], Fe$_3$O$_4$ [586], ZnO [587], and Ag [588, 589] spherical nanoparticles were used in the literature. One advantage of TiO$_2$ nanoparticles is their photocatalytic activity, which is interesting in photolithography. The morphology of the nanoparticles can also be easily controlled with the synthesis parameters [580–583]. Robust superoleophobic surfaces ($\theta_{hexadecane} = 154.7°$, $H_{hexadecane} = 21.8°$ and $\alpha_{hexadecane} = 18.9°$) with multiscale roughness were obtained by spraying copper perfluorooctanoate suspension [590, 591].

This method is very interesting because many metals and carboxylates can be used. Lei et al. also developed a one-step spray deposition of metal stearate particles [592]. The color of the substrate could be easily controlled with the use metal (zinc, chromium, cobalt, iron, and copper). The substrates displayed also superhydrophobic properties with excellent corrosion resistance.

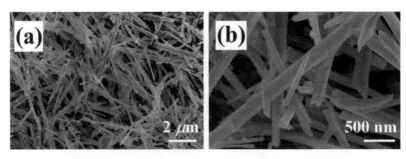

Figure 85 Sb_2S_3 micro-/nanorods obtained with the polyol process. Reprinted with permission from ref. 597, copyright 2014, Elsevier.

Silver nanowires were produced using the polyol process by reduction of silver ions by ethylene glycol at high temperature and by microwave [593]. The height of the nanowires could be controlled using different treatment times. The longest silver nanowires gave rise to the highest oleophobic properties ($\theta_{\text{ethylene glycol}}$ = 146.2°, $H_{\text{ethylene glycol}}$ = 4.3°, $\alpha_{\text{ethylene glycol}}$ ≈ 18°). Moreover, these silver materials can be used as antibacterial coating and for SERS [594]. Bioinspired by mussel adhesives, it is also possible to coat on various substrates and with an adherence using polydopamine [595]. Indeed, polydopamine allows the grafting on both the substrate and the nanoparticles. $In(OH)_3$ nanoparticles were also coated on PDMS sponge using polydopamine [596]. Moreover, the sponge could detect ammonia by switching its surface wettability. Chen et al. also reported the formation of Sb_2S_3 micro/nanorods via a rapid microwave synthesis in ethylene glycol and the presence of $SbCl_3$, $Na_2S_2O_3$, and thiosemicarbazide [597] (Figure 85). Depending on the diameter of the Sb_2S_3 micro/nanorods, surface properties ranging from superhydrophilic to superhydrophobic were obtained. Petal-like $Ni_xCd_{1-x}S$ nanostructures modified with octadecylamine were reported using a facile liquid–liquid interfacial self-assembly [598]. The substrates displayed both superhydrophobic properties and photoluminescence. Jiang et al. developed ZnO hollow microspheres with controlled architectures, including 1D nanowires, 2D nanosheets, and 3D mesoporous nanoball blocks, via a two-step self-assembly process in the presence of zinc acetate, ethanol, hexamethylenetetramine, and Pluronic P123, as surfactant [599]. The substrates displayed robust superhydrophobic properties with long-term UV resistance.

Carbon nanoparticles, including multiwalled carbon nanotubes, carbon fibers, and diamond-like carbon, were also employed in the literature [600–602]. In order to obtain superoleophobic properties with carbon nanoparticles, a substrate can be held in the flame of a candle to deposit a soot layer [603–614]. Here, the superoleophobic properties ($\theta_{hexadecane}$ = 156° and $\alpha_{hexadecane}$ = 5°) could be enhanced by a CVD of SiO_2 nanoparticles or by annealing above 1100°C, which resulted in the formation of more rod-like nanostructures. The size of the particles can be controlled by the distance from the flame. The optimal contact angles were obtained after sooting in the middle of the flame. Due to the possibility of easily moving droplets on these substrates, they were investigated in digital microfluidics [612] (Figure 86). Mixtures of carbon nanotubes and SiO_2 or TiO_2 nanoparticles were also used in the literature [615].

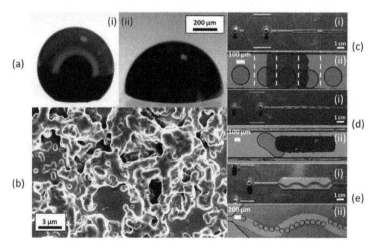

Figure 86 Candle soot deposition by applying the flame of a candle on a substrate. The substrates could be used for the control of liquid droplet. Reprinted with permission from ref. 612, copyright 2013, American Chemical Society.

3.9.2 Nanocomposites

In order to enhance the mechanical, adherence and durability properties of the coatings obtained by deposition of nanoparticles, a polymer pre-coating can be used before adding the nanoparticles. Otherwise, one of the most used methods is to prepare nanocomposites. Different strategies have been developed to obtain nanocomposites. For example, the polymer can be introduced during the nanoparticles synthesis, for example, by adding TEOS to form SiO_2/polymer nanocomposites [616–618]. Using this process, a perfluoroether containing two terminal carboxylic acid groups was used in the presence of TEOS [617]. Surprisingly, the resulting materials displayed both superoleophobic ($\theta_{dodecane}$ = 129°) and superhydrophilic properties, which is extremely rare in the literature. Durable superhydrophobic ZnO/PTFE films were also developed by introducing zinc acetate and NaCl in a solution containing PTFE [618].

The polymer can also be introduced after the nanoparticle synthesis. Different polymers, including polyurethanes, polypropylene (PP), polystyrene (PS), and polyvinyl butyral (PVB), were used to obtain superhydrophobic properties [619–623] (Figure 87). Spray of polyurethane/molybdenum disulfide (MoS_2) was used to obtain coatings with wear resistance. After modification with a fluorinated silane, the coatings showed superoleophobic properties ($\theta_{hexadecane}$ = 151° and $H_{hexadecane}$ = 30°) [625]. Tiemblo et al. also developed fluorescent coating by mixing SiO_2 nanoparticles with a highly fluorescent and hydrophobic conducting polymer (poly(9,9-dioctyl-9H-fluorene)) [624]. Other groups used also quantum dots or fluorescent organic molecules (europium compounds) to obtain fluorescent coatings [626]. Polydimethylsiloxane (PDMS) was also used by numerous research groups for its high elasticity and high hydrophobic properties [627–636]. For example, nanocomposites made by mixing TiO_2 nanowires and PDMS displayed superhydrophobic properties with switchable wettability after UV irradiation. Multiwalled carbon nanotubes (MWCNTs) and Ag nanostructures were also used to obtain conductive coatings [630]. Superoleophobic nanoparticles/PDMS nanocomposites can also be produced by modifying the extreme surface with a fluorinated monolayer [633, 634]. This is possible by creating hydroxyl functions in

PDMS backbone by immersing the nanocomposite substrate in piranha solution or by plasma treatment, for example.

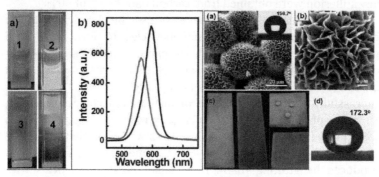

Figure 87 Superhydrophobic and fluorescent SiO$_2$/polyurethane nanocomposites after incorporation of CdTe quantum dots. Reprinted with permission from ref. 625, copyright 2010, American Chemical Society.

Otherwise, fluoropolymers can also be used to obtain superoleophobic properties [637–641]. For example, superoleophobic properties with low oil adhesion were obtained ($\theta_{hexadecane}$ = 158° and $\alpha_{hexadecane}$ = 5.1°) by mixing fluorinated SiO$_2$ and poly(vinylidene fluoride-hexafluoro-propylene) [637]. Mabry et al. also used fluorinated SiO$_2$ and fluoropolymer Viton ETP-600S (DuPont) to obtain superoleophobic properties [638]. However, the authors observed that rather than being present between the interstices between the nanoparticles, the nanoparticles were widely distributed across the surface roughness. Superoleophobic properties ($\theta_{hexadecane}$ = 152° and $\alpha_{hexadecane}$ = 40°) were also obtained by mixing fluorinated MWCNTs with fluorinated polyurethane [640]. In order to enhance the substrate adherence, fluorinated polyacrylates were also used [642–650]. Hsieh et al. mixed fluorinated polyacrylates to 20 nm SiO$_2$ nanoparticles and obtained superoleophobic properties ($\theta_{ethylene\ glycol}$ = 165.2° and $H_{ethylene\ glycol}$ = 2.5°) for a F/Si ratio of 2.13 [642, 643]. Steel et al. used fluorinated polyacrylates and 50 nm ZnO nanoparticles and obtained the highest oleophobic properties ($\theta_{hexadecane}$ = 154° and $H_{hexadecane}$ = 6°) for a ZnO:polyacrylates mass fraction of 3.3 [644]. Superoleophobic and conductive coatings were also obtained by mixing fluorinated polyacrylates and carbon nanofibers (≈100 nm in diameter and ≈130 μm in length) [645]. Both the superoleophobic properties and electrical conductivity could be controlled by carbon nanofibers content. Superoleophobic properties with low adhesion were obtained from carbon nanofibers of 60%.

The mechanical properties can also be highly enhanced by grafting polymer on the nanoparticles to form core-shell particles [651–656]. Liu et al. prepared a diblock copolymer containing both fluorinated chains and triisopropyloxysilyl groups for the grafting on SiO_2 nanoparticles [651]. The authors successfully coated the surface of SiO_2 nanoparticles uniformly. Superoleophobic properties ($\theta_{hexadecane}$ = 149° and $H_{hexadecane}$ = 13°) were measured using these materials as well as a high resistance in a basic solution. Raspberry-like polymer particles were also produced by grafting [655]. Here, small polymer particles with glycidyl groups were mixed to larger polymer particles containing amine groups. A fluropolymer with multiple triethoxyxilane groups was also used to graft on ZnO nanostructures [656] (Figure 88).

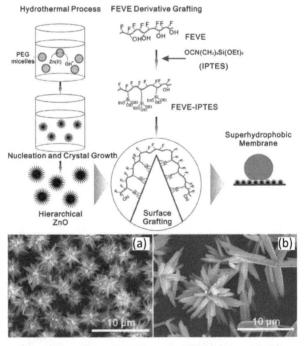

Figure 88 Superhydrophobic ZnO/fluoropolymer nanocomposites using a grafting strategy. Reprinted with permission from ref. 656, copyright 2015, Elsevier.

Cho et al. reported the preparation of SiO$_2$ nanoparticles with methacrylate groups [657]. Then, a fluorinated methacrylate was added to obtain superoleophobic properties. Wang et al. modified multiwalled carbon nanotubes with bromide groups and then a polymer containing ammonium groups was initiated on the surface of nanotubes by ATRP [658, 659]. The resulting materials displayed reversible wettability from superoleophobic and superoleophilic by successive anion exchanges using thiocyanate and perfluorooctanoate anions. Patton et al. mixed SiO$_2$ nanoparticles with trimethylsilyl groups with a formulation containing a photoinitiator a trifunctional alkene and fluorinated thiols [660, 661] (Figure 89). After spray coating in the presence of a UV lamp, superoleophobic thiol-ene resins ($\theta_{hexadecane}$ = 155.3°, $H_{hexadecane}$ = 9.5° and $\alpha_{hexadecane}$ = 4° using 30 wt% of SiO$_2$) were obtained. Following this strategy, a polyacrylate resin obtained by photopolymerization was also used [662].

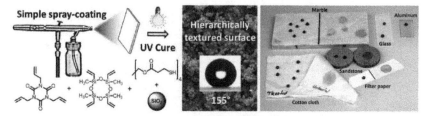

Figure 89 Superhydrophobic photopolymerizable nanocomposited obtained by spray coating in the presence of a UV lamp. Reprinted with permission from ref. 661, copyright 2013, American Chemical Society.

Another strategy used in the literature was to coat nanoparticles with polymer using a layer-by-layer (LbL) approach [663–667]. For example, TiO$_2$ nanoparticles were coated with successive layer of poly(acrylic acid) and a perfluoroalkyl methacrylic copolymer to obtain superoleophobic properties [664]. Kang et al. deposited on SiO$_2$ nanoparticles positively charged multiwall nanotubes and negatively charged reduced graphene oxide [665]. Carbon nanotubes were also modified using a LbL approach using polydopamine and poly(ethyleneimine) [666, 667].

3.9.3 Colloidal Lithography

In order to prepare extremely ordered silica nanospheres on substrate, colloidal lithography can be used [670–674]. In this process, SiO_2 spheres having the same diameter are spin-coated or deposited on substrates using the Langmuir–Blodgett technique, to produce close-packed spheres (Figure 90). Large spheres usually ≥100 nm are used in the technique.

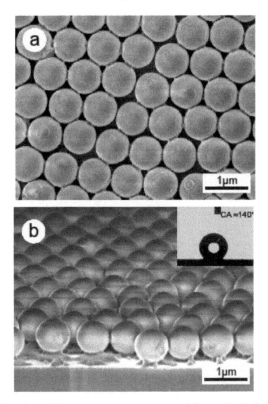

Figure 90 Stacked SiO_2 nanospheres obtained by colloidal lithography. Reprinted with permission from ref. 670, copyright 2010, American Chemical Society.

Kooij et al. developed close-packed SiO_2 sphere arrays of different size (140, 440, and 830 nm) and modified them using a fluorinated silane [670]. They obtained superhydrophobic properties with very low α_w irrespective of the size of SiO_2 spheres. They also observed that their superhydrophobic behavior is also due to the formation of siloxane nanostructures on the SiO$_2$ spheres. Indeed, one of the easiest ways to produce stable superhydrophobic properties is to induce a dual-scale roughness by forming nanoroughness on the SiO$_2$ spheres. For example, Hsieh et al. studied not only SiO$_2$ spheres of small and large size but also mixing of SiO$_2$ spheres of two different sizes [671]. After modification with a fluorinated copolymer, the highest superoleophobic properties ($\theta_{\text{isopropanol}}$ = 148.6°) were obtained after mixing SiO$_2$ spheres of two different sizes. Highly ordered raspberry-like structures with very low hysteresis were also reported in the literature using two different strategies. Jiang et al. used a second colloidal lithography process by using very small SiO$_2$ nanospheres to obtain raspberry-like structure arrays [673] (Figure 91). By contrast, Lee et al. grafted small nanoparticles on the surface of SiO$_2$ nanospheres [674].

Figure 91 Raspberry-like SiO$_2$ structures obtained by a two-step colloidal lithography. Reprinted with permission from ref. 673, copyright 2013, Elsevier.

Many other structured materials were used to induce nanoroughness on SiO$_2$ nanospheres [675–679]. For example, several research groups used gold nanoparticles [675–677]. Cho et al. deposited using a layer-by-layer process positively charged and negatively charged nanometer-size block copolymer micelles to produce raspberry-like structures [678]. The surface hydrophobicity could be easily controlled with the nanoscale roughness, which is linked to the Mw of the block segments and the charge density of the hydrophilic corona blocks.

The surface morphology can also be controlled by etching the SiO$_2$ nanospheres [680–682]. Nanotube arrays were developed by RIE. The substrates displayed both superhydrophobic properties and broadband antireflective properties and could be used in high-density data storage and optoelectronics to biological sensing and subwavelength optics (Figure 92).

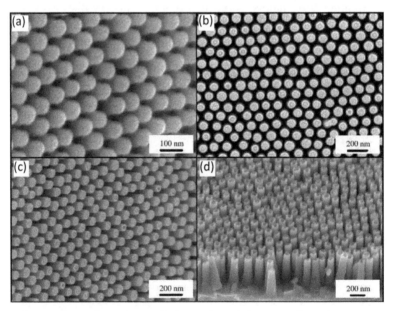

Figure 92 Nanopillars obtained after etching SiO$_2$ nanospheres by reactive ion etching. Reprinted with permission from ref. 680, copyright 2008, IOP Publishing.

3.9.4 Textured Substrates

Textured substrates such as fabrics, membranes, sponges, foams, and meshes can also be modified with nanoparticles or nanocomposites in order to enhance the range of applications [683–697]. Fluorinated SiO$_2$ nanoparticles were used on fabrics in order to obtain self-cleaning textiles for various liquids and particles. Lin et al. developed pH-switchable fabrics by coating with SiO$_2$ nanoparticles and decanoic acid-modified TiO$_2$ nanoparticles [685]. These selective fabrics could be used to separate oil/water mixtures. Textiles with antibacterial properties were also reported using Ag nanoparticles [686, 687]. Superhydrophobic and superoleophilic sponges with high oil-adsorption properties were prepared by modifying polyurethane sponges with ZnO microrods and palmitic acid [694, 695]. Stainless steel meshes were also modified with ZnO nanoparticles treated with stearic acid [696, 697]. The resulting meshes showed photo-induced switchable wettability and could be used for on-demand separation of oil/water mixtures.

3.9.5 Aerogels

3.9.5.1 Silica aerogels

Other materials can be prepared using the sol-gel process. Among them, xerogels and aerogels are a unique class of materials. Indeed, silica aerogels are nanoporous, have high surface area, extremely low density, low thermal conductivity, low dielectric constant, and low index of refraction. Typically, drying the wet gel under ambient conditions results in xerogels (60–90%), whereas under supercritical conditions in aerogels (90–99% air). However, the inherent fragility of the aerogels made from TEOS or tetramethoxysilane (TMOS) due to their highly hygroscopic behavior. Hence, strategies have been developed to reinforce these aerogels and to suppress the hydroscopic behavior.

In order to suppress the inherent hydroscopic behavior, organosilanes such as methyltriethoxysilane (MTES), methyltrimethoxysilane (MTMS) or hexamethyl disilazane (HMDS) are used [698–713] (Figure 93). For example, flexible and thermally stable superhydrophobic (θ_w) aerogels were prepared using MTES in ethanol and using HCl or oxalic acid and NH$_4$OH. Superhydrophobic and superoleophilic "sponge-like" aerogels were also reported in a very high adsorption capacity for various kinds or organic liquids by mixing MTES and dimethyldiethoxysilane (DMDES) [702]. These materials can be used to remove oils from water. Carroll et al. used a mixture of TMOS and MTMS and obtained superhydrophobic aerogels for a MTMS percentage in the range 10–75% [703]. Several other authors reported HMDS as hydrophobic organosilane. Poly(methylhydrosiloxane) (PHMS) was also used in the literature to obtain superhydrophobic aerogels [710–713]. In order to prepare superoleophobic silica aerogels with resistance to mechanical damage, the aerogels were coated with a surfactant which allows self-replenishing [712].

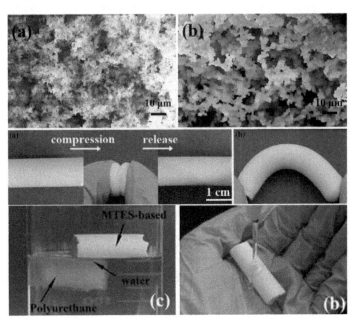

Figure 93 Superhydrophobic silica aerogels obtained in the presence of MTES. Reprinted with permission from ref. 701, copyright 2014, Springer.

Moreover, the aerogels can be reinforced with a polymer [714–718]. Polystyrene/silica aerogels with fast absorption capacity for various hydrocarbon liquids were reported [714]. PDMS modified with hydroxyl groups were also reported in the literature [717].

3.9.5.2 Graphene aerogels

Superhydrophobic graphene aerogels were also studied in the literature (Figure 94). These aerogels are extremely interesting for their electrical conductivity and can be used as electrodes for supercapacitors and lithium ion batteries. Graphene oxide aerogels were obtained using Hummers' method. Typically, aqueous dispersions of graphene oxide are prepared by sonication. Then, the solution is frozen using methanol and dry ice and freeze-dried to form the graphene oxide aerogel. Finally, the aerogel is thermally reduced in argon at very high temperature (>1000°C) and with a low-temperature ramp in order to avoid material deflagration. During the thermal reduction, the morphology and the open pore structure of the aerogel are preserved, while the hydrophilic groups (hydroxyls, carbonyls, carboxylic acids, epoxides, etc.) are reduced rendering the surface hydrophobic. Using this method, Lin et al. obtained the best results (θ_w = 160°) for a graphene oxide concentration of 7 mg L^{-1} [719]. Different compounds can be incorporated using this method [720].

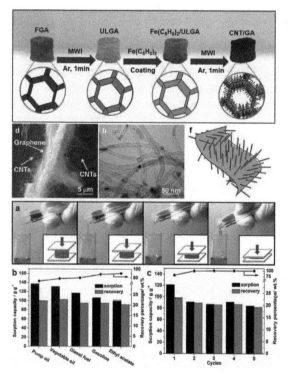

Figure 94 Superhydrophobic multiwalled carbon nanotube–graphene aerogels. The materials could be used for oil sorption. Reprinted with permission from ref. 725, copyright 2014, American Chemical Society.

Superhydrophobic and superoleophilic PVDF/graphene oxide were prepared by mixing graphene oxide and PVDF [721]. The materials showed also high absorption capacity for oils and organic solvents, excellent absorption recyclability and considerable mechanical properties. They can be used for oil/water separation, oil spill cleanup, and recovery of organic solvents. Ellison et al. reported the preparation of superhydrophobic poly(acrylic acid)/graphene oxide aerogels with high elasticity and mechanical properties [722]. PDMS/graphene oxide aerogels were also reported in the literature [723]. The materials displayed high compressibility and also high water adhesion due to their foamlike surface structure. Li et al. also developed TiO_2/graphene oxide aerogels [724]. The resulting materials displayed tunable surface wettability with photocatalytic properties due to the presence of TiO_2.

Other research groups also worked on the process of graphene oxide aerogel formation [725–729]. Jiang et al. used (3-mercaptopropyl)trimethoxysilane (MPS) to facilitate the formation of the aerogel [729]. The role of MPS is to form silanol groups, which complete the reduction process at low temperature (100°C) and also create space between graphene sheets by in situ growth of silicon polymer. The materials displayed stable superhydrophobic properties ($\theta_w > 160°$) upon washing–drying cycles and exceptional thermal stability. Khor et al. used the Spark Plasma Sintering process to obtain graphene oxide aerogel [728]. In this process, pulsed direct current flows through both the die and the sample generate spark plasma within the sample. Here, graphene sheet exfoliation was obtained at 500°C and superhydrophobic properties were obtained after heating for 1 min at 1050°C.

Some other materials were used to produce superhydrophobic aerogels. Carbon fiber aerogels were produced by dipping cotton pieces in $Fe(NO_3)_3$ solution, pyrolyzing under argon at a low-temperature ramp and down cooled down to room temperature [730]. The resulting cotton displayed both superhydrophobic and superoleophilic properties and could be used for oil/water separation. Zettl et al. produced superhydrophobic ($\theta_w = 155°$) boron nitride (BN) aerogels with high oil absorption capacity by heating BN powder to 1600–1800°C by a radio frequency induction furnace under 2000 sccm flow of N_2 gas [731].

3.10 Chemical Vapor Deposition

3.10.1 Carbon

In chemical vapor deposition, a volatile carbon source such as acetylene, ethylene, methane, xylene, or ethanol is decomposed at very high temperature and in the presence of a catalysis: for example, with acetylene $C_2H_2(g) \rightarrow 2C(s) + H_2(g)$. Using this process, carbon nanotubes can be produced and their parameters depend, for example, on the carbon source, the gas mixture flux, the level of impurities and the necessary catalyst (for example, Ni, Fe, Co and their bimetallic alloys).

Several research groups reported that vertically aligned MWCNTs displayed extremely high adhesion, even higher that the adhesion force of natural gecko foot, and can be used as dry adhesives [732–735] (Figures 95 and 96).

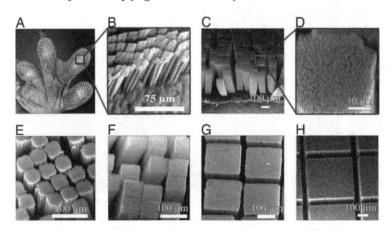

Figure 95 Preparation of carbon nanotube-based synthetic gecko tapes. Reprinted with permission from ref. 733, copyright 2007, Proceedings of the National Academy of Sciences of the United States of America.

Even if carbon is an intrinsically hydrophilic material ($\theta_W^Y \approx 84°$), by controlling the areal density, the diameter, and the height of the nanotubes with the experimental conditions, it is possible to achieve superhydrophobic properties ($\theta_W > 150°$) and low adhesion [732–739]. It is possible to induce the growth of these materials on various substrates even on complex geometry by deposing a thin layer of a metal such as nickel used as catalyst

[736–738]. It was also shown that vertically aligned MWCNTs can cause considerable drag reduction due to a slip length as low as 10 µm [739]. It is also possible to introduce oxygen groups on MWCNTs and, as a consequence, to decrease the hydrophobicity by oxygen plasma and UV/ozone and later to increase it by vacuum annealing or by CO_2 laser treatment [740–742].

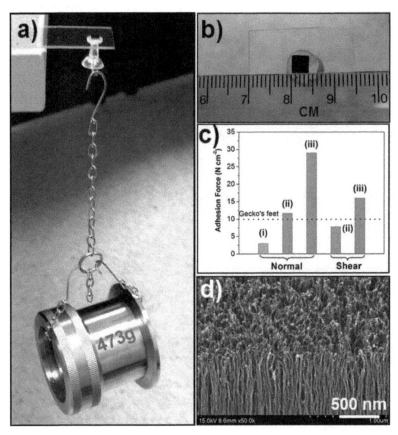

Figure 96 Multiwalled carbon nanotube substrates with high adhesive forces. Reprinted with permission from ref. 735, copyright 2007, Wiley.

Star-shaped carbon nanotubes with superhydrophobic properties (θ_w = 165°) could also be produced by pulse laser deposition of a nanocomposite film before the growth of carbon nanotubes [743]. Here, the nanocomposite film consists of metal catalysts embedded into a dielectric matrix such as diamond-like carbon or AlN. The role of the dielectric matrix is to inhibit the growth of the metal and to form metal nanospheres. These materials are excellent candidates for solar thermal harvesting due to their extremely low optical reflection (reflectivity < 10^{-4}).

Because vertically aligned MWCNTs have nanoscale roughness and carbon is a material of relatively high surface energy, they are often not sufficient to lead to superhydrophobic properties with low adhesion. This could be achieved by creating vertically aligned double-layered carbon nanotubes [744]. Otherwise, fluorinated materials are often used to coat MWCNTs. Zhu and coworkers used fluoroalkyl silane to modify MWCNTs and the resulting substrates could impede the wetting by many kinds of liquid even of low surface tension [745]. Gleason et al. coated MWCNTs with an ultra-thin layer of PTFE to coat carbon nanotube arrays with an areal density of 10 per μm^2, a mean diameter of 50 nm, and height of 2 µm [746]. On these substrates, water droplets could easily bounce indicating low water adhesion (H_w = 10°). Moreover, the authors showed that the receding contact angles is very sensitive to differences in nanotube height but this sensitivity is much lower after PTFE coating. MWCNTs can also be treated with different chemicals in order to obtain other functionalities. For example, after functionalization with silicone, it was possible to reversely change the surface wettability by annealing and corona discharge [747]. MWCNTs were coated by ZnO thin layer and the resulting substrates displayed reversible change from superhydrophobicity to hydrophilicity by alternating UV irradiation and dark storage [748]. Hu et al. covalently functionalized MWCNTs with polyhedral oligomeric silsequioxane (POSS) and observed that the resulting substrates possessed also flame-retardant properties [749]. In order to prepare electrode materials, graphene oxides were grafted on MWCNTs. For that, MWCNTs were first exfoliated and functionalized with oxygen groups by using plasma etching. The electrodes displayed fast electron transfer kinetics and are excellent candidates for electrocatalysis and biosensors [750, 751]. For example, these electrodes were used to determine at nanomolar levels atorvastatin calcium in pharmaceutical and biological samples (Figure 97).

100 | Fabrication Processes

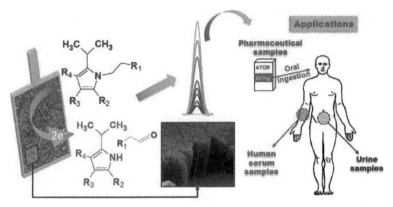

Figure 97 Superhydrophobic multiwalled carbon nanotubes used to determine at nanomolar levels atorvastatin calcium in pharmaceutical and biological samples. Reprinted with permission from ref. 751, copyright 2014, Royal Society of Chemistry.

Finally, the robustness of the superhydrophobic properties can also be enhanced by growing MWCNTs on microstructured substrates. Kim et al. studied the growth of MWCNTs coated with silicone on micropillar arrays with pillar-to-pillar spacings ranging from 45 to 160 μm and a diameter of 65 μm [752] (Figure 98). The presence of the micropatterns enhanced the superhydrophobic properties with θ_w = 168°, H_w = 2.64° and α_w < 5° for a pillar-to-pillar spacing of 160 μm. Moreover, the substrates resist the bouncing of water droplets for impact velocity up to ≈ 1.4 m/s, which is due to their extremely high antiwetting capillary pressure. Jiang et al. prepared micropatterned substrates of different pillar spacing by photolithography on which MWCNTs were grown [753]. After modification with a fluorinated silane, they observed superhydrophobic properties with θ_w > 154° irrespective of the pillar spacing. By contrast, after modification with vinyltrimethoxysilane, the substrates displayed huge hydrophobicity differences from 20.8° to 154.9° as the pillar spacing changes [754]. Other research groups also studied the growth of carbon nanofibers on smooth substrates and on fabrics by coating first with Ni-doped mesoporous silica [755, 756]. Qiu et al. first deposited on dendritic copper structures by electrodeposition or galvanic replacement to produce highly rough substrates while the copper was used as catalyst for the growth of carbon nanofibers [757]. The authors showed that these substrates can be used as slippery liquid-infused porous surface and could be used as anti-icing and anticorrosion coatings.

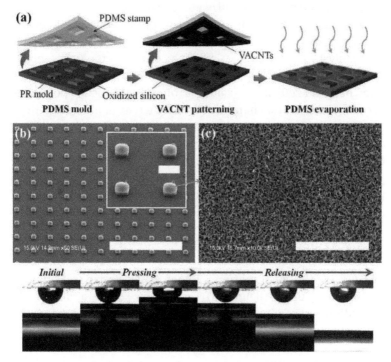

Figure 98 Superhydrophobic surfaces with low adhesion by growth of multiwalled carbon nanotubes on micropillar arrays. Reprinted with permission from ref. 752, copyright 2014, Elsevier.

Membranes based on aligned MWCNTs were fabricated and studied by several research groups [758–762]. The membranes enhanced vapor condensation, selectively rejected water vapor while allowing a fast transport of dry gases. Moreover, these properties also depended on the membrane temperature. Furthermore, the membranes also displayed high selectivity. These membranes can also be used for water desalination. SiC membranes were also modified by MWCNTs to obtain superhydrophobic and superoleophilic properties [763, 764]. These membranes could be used for oil/water separation.

Other superhydrophobic carbon-based materials were also reported by CVD. For example, diamond-like carbon (a more hydrophilic carbon-based material) rough films were reported by microwave plasma CVD [765, 766]. The growth of diamond-like carbon was possible on textile substrates and the resulting materials could be used for oil/water separation.

Graphene nanosheets were also produced by microwave plasma CVD at a relatively low temperature and using copper as catalyst [767, 768]. Graphene nanosheets were highly hydrophobic with θ_w = 132.9°. The graphene nanosheets could also be oxidized in a solution of sulfuric acid, potassium permanganate and sodium nitrate in order to functionalize them by carboxylic acids and reduce the surface hydrophobicity. Then, octadecylamine was used to react with the carboxylic acids and to increase the surface hydrophobicity. Moreover, the surface properties could also be enhanced by substrate pre-patterning [769]. Superhydrophobic graphene foams were also prepared using sacrificial nickel foam [770].

3.10.2 Silicon

By changing the chemical source, other materials can be produced. For example, using silane (SiH_4) as silicon source and after the deposition on an ultra-thin gold layer as catalyst, aligned silicon nanowires were reported. The reaction is

$$SiH_4 (g) \rightarrow Si(s) + 2H_2(g).$$

The Si nanowires are especially composed of nanowires of two different lengths (double nanotextures), while these lengths also depend on the thickness of the gold layer [771–773]. After electrically insulating with 300 nm SiO_2 and hydrophobization with a fluoropolymer, the substrates displayed superhydrophobic properties ($\theta_w \approx 160°$) with ultra-low hysteresis ($H < 1°$). Moreover, these substrates could be used for electrowetting experiments with a maximum contact angle variation of 35° at 190 V_{TRMS}. In order to develop superoleophobic properties, Si nanowires were also grown on microgrooves with various re-entrant structures on the sidewalls (Figure 99). The best results ($\theta_{hexadecane} \approx 140°$) were obtained with a double layer of long nanowires.

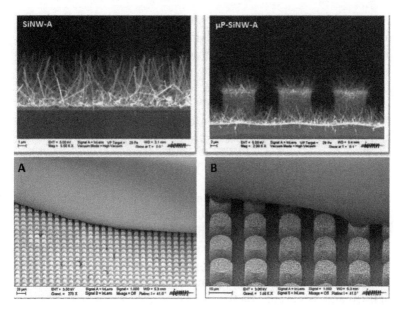

Figure 99 Superoleophobic surfaces with low adhesion by growth of silicon nanowires on micropillar arrays. Reprinted with permission from ref. 773, copyright 2014, Elsevier.

3.10.3 Carbides and Nitrides

To obtain many kinds of materials, different chemical sources can be introduced or only one chemical source can introduce several different atoms. For example, SiC nanorods were obtained by introducing hexamethyldisilane in H_2 flow at high temperature in an atmospheric pressure CVD (APCVD) process [774]. Hence, SiC can be formed following this reaction:

$$C_6H_{18}Si_2(g) \rightarrow 2SiC(s) + 4\,CH_4(g) + 2\,H_2(g)$$

By using CH_4 and N_2 as carbon and nitrogen source, respectively, and H_2 as gas flow, carbon nitride (CN_x) was prepared by plasma-enhanced CVD [775]. CN_x is an intrinsically hydrophilic material with θ_w^Y = 53.5°, but using an electrode distance of 1 cm, it was possible to obtain fibrous nanostructures with superhydrophobic properties (θ_w = 154.6°).

Several nitrides such as BN, AlN, and GaN are semiconductors and are extensively used in photonics and electronics applications such as deep-ultraviolet light-emitting diodes (LEDs), field emitters, surface acoustic wave device or ultraviolet sensors. Different research groups were interested in the formation of boron nitride (BN) nanostructures by CVD [776–781]. By mixing B as well as MeO and FeO as oxidants or boron trioxide (B_2O_3) and using NH_3 as gas flow, BN nanosheets were obtained:

$$(B_2O_3(g) + 2\ NH_3(g) \to 2BN(s) + 3\ H_2O(g))$$

BN nanosheets were also reported using BF_3, N_2, and H_2 as gas mixture. At high temperature (1200°C) extremely rough substrates were obtained with superhydrophobic properties (θ_w = 159.1°) with θ_w^Y = 51.4°. Moreover, it was also possible to control their surface hydrophobicity by air-plasma treatment [780]. The authors showed that such treatment could induce the formation of B-OH and N-H groups on the surface.

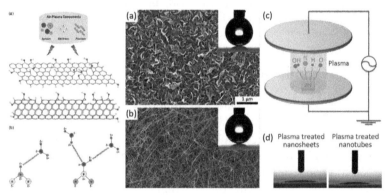

Figure 100 Superhydrophobic surfaces obtained by growth of BN nanosheets. Reprinted with permission from ref. 780, copyright 2014, American Chemical Society.

Aluminum nitride (AlN) was prepared at 1310°C using Al powder, W as catalyst and N_2 as gas flow [782]. After 120 min, high-density AlN nanowires were observed. The substrates were superhydrophilic but after few weeks in air at room temperature, the substrates displayed superhydrophobic properties (θ_w > 150°, α_w < 10°). Moreover, the surface wettability is reversible from superhydrophobic to superhydrophilic by UV irradiation.

High-density GaN nanowires were also prepared using gallium, gallium oxide, and graphite. Using N_2 as gas flow and at 960°C, GaN nanowires were observed after 30 min [783]. The as-prepared GaN nanowires were superhydrophobic (θ_w = 155°), became superhydrophilic after UV irradiation and became superhydrophobic again after dark storage or heating. Moreover, the authors also observed the possibility of modulating protein and cell adhesion by UV irradiation.

3.10.4 Oxides and Sulfides

Many research groups have been also interested in the formation of nanostructured oxides and sulfides by CVD. ZnO nanorods were reported using zinc powder and using argon and oxygen as gas flow [784]. In order to obtain surfaces with micro- and nanostructures, these nanorods were also grown on anodized aluminum oxide (Figure 101). The resulting substrates were parahydrophobic with θ_w = 151° and had extremely high adhesion. Other groups reported the decoration of these nanorods by Ag nanoparticles [785–787]. These substrates are extremely interesting for SERS studies. ZnS nanorods were also obtained without oxygen but by adding sulfur powder and graphite as catalyst [788].

Figure 101 Parahydrophobic surfaces obtained by growth of ZnO nanorods on anodized aluminum oxide. Reprinted with permission from ref. 784, copyright 2008, American Chemical Society.

A similar approach in the literature was used to develop indium oxide (In_2O_3), vanadium oxide (V_2O_5), and tungsten oxide (WO_x) nanowires [789–792] (Figure 102).

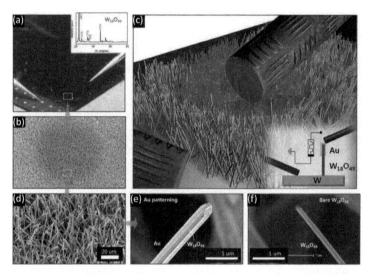

Figure 102 Superhydrophobic WO$_x$-Au core–shell nanowires with switchable wettability and high stability underwater. Reprinted with permission from ref. 792, copyright 2012, Wiley.

Singh et al. reported the formation of randomly distributed and vertically aligned In$_2$O$_3$ nanorods with parahydrophobic (θ_w = 133.7°) and superhydrophobic (θ_w = 159.3°) properties, respectively [793] (Figure 103). The superhydrophobic properties of vertically aligned nanorods can be easily explained by the low solid–liquid fraction when a water droplet is sit on top of the nanorods. By contrast, Ma et al. used InP and GaP substrates to induce the growth by CVD of superhydrophobic In$_2$O$_3$ nanowires and β-Ga$_2$O$_3$ nanowires [794, 795]. By using tetracyanoquinodimethane (TNCQ) and copper substrates, superhydrophobic CuTCNQ nanowires were obtained [796]. Moreover, the substrates displayed high adhesion of osteoblast.

Xiong et al. reported the formation of tin oxide SnO$_2$ nanowires by CVD by decomposition of Sn(O-tBu)$_4$ at 650–750°C [797]. These substrates were superhydrophilic. However, it was possible to induce the growth of SnO$_2$ or VO$_x$ (using VO(O-iPr)$_3$) from SnO$_2$ nanowires using a second CVD step, which allows reaching θ_w = 133.0° and even θ_w = 155.8° after three CVD steps. Moreover, their surface wettability could be reversely switch from superhydrophobic to superhydrophilic by UV irradiation and dark storage or O$_2$ annealing.

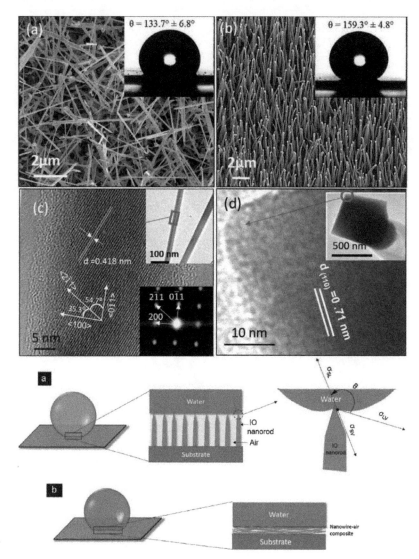

Figure 103 Superhydrophobic SnO$_2$@SnO$_2$ heterostructures with switchable wettability. Reprinted with permission from ref. 797, copyright 2011, American Chemical Society.

Finally, superhydrophobic nanostructured SiO$_2$ films were also reported by atmospheric pressure plasma-enhanced chemical vapor deposition using tetrakis(trimethylsilyloxy)silane as silicon source [798, 799].

Chapter 4

Conclusion

In this book, we focused on fabrication processes to obtain superhydrophobic and related properties, especially using metallic and inorganic materials. Indeed, these materials present unique properties, for example, in terms of thermal resistance, mechanical resistance, chemical and ageing resistance, optical (transparency, antireflection, photoluminescence), and electrical properties (conducting, semi-conducting, insulating). We showed that these materials are especially produced using different strategies: etching in acidic media, plasma processes, laser treatments, etching in basic media, anodization, electrodeposition, electroless deposition, hydrothermal processes, use of nanoparticles, and chemical vapor deposition. This book has allowed us to bring together the strategies used in the literature by researchers and will help find new strategies to develop such kind of materials with superhydrophobic properties.

Bioinspired Superhydrophobic Surfaces: Advances and Applications with Metallic and Inorganic Materials
Frédéric Guittard and Thierry Darmanin
Copyright © 2018 Pan Stanford Publishing Pte. Ltd.
ISBN 978-981-4774-05-5 (Hardcover), 978-1-315-22961-4 (eBook)
www.panstanford.com

References

1. Darmanin, T., and Guittard, F. (2015). Superhydrophobic and superoleophobic properties in nature, *Mater. Today*, **18**, pp. 273–285.
2. Koch, K., Bhushan, B., and Barthlott, W. (2009). Multifunctional surface structures of plants: An inspiration for biomimetics, *Prog. Mater. Sci.*, **54**, pp. 137–178.
3. Koch, K., Bhushan, B., and Barthlott, W. (2008). Diversity of structure, morphology and wetting of plant surfaces, *Soft Matter*, **4**, pp. 1943–1963.
4. Afsal, M., and Chen, L.-J. (2011). Anomalous adhesive superhydrophobicity on aligned ZnO nanowire arrays grown on a lotus leaf, *J. Mater. Chem.*, **21**, pp. 18061–18066.
5. Lepore, E., Giorcelli, M., Saggese, C., Tagliaferro, A., and Pugno, N. (2013). Mimicking water striders' legs superhydrophobicity and buoyancy with cabbage leaves and nanotube carpets, *J. Mater. Res.*, **28**, pp. 976–983.
6. Hamlett, C. A. E., Shirtcliffe, N. J., Pyatt, F. B., Newton, M. I., McHale, G., and Koch, K. (2011). Passive water control at the surface of a superhydrophobic lichen, *Planta*, **234**, pp. 1267–1274.
7. Watson, G. S., Cribb, B. W., and Watson, J. A. (2010). The role of micro/nano channel structuring in repelling water on cuticle arrays of the lacewing, *J. Struct. Biol.*, **171**, pp. 44–51.
8. Sun, Z., Liao, T., Liu, K., Jiang, L., Kim, J. H., and Dou, S. X. (2014). Fly-eye inspired superhydrophobic anti-fogging inorganic nanostructures, *Small*, **10**, pp. 3001–3006.
9. Gao, X., Yan, X., Yao, X., Xu, L., Zhang, K., Zhang, J., Yang, B., and Jiang, L. (2007). The dry-style antifogging properties of mosquito compound eyes and artificial analogues prepared by soft lithography, *Adv. Mater.*, **19**, pp. 2213–2217.
10. Bush, J. W. M., Hu, D. L., and Prakash, M. (2007). The integument of water-walking arthropods: Form and function, *Adv. Insect Physiol.*, **34**, pp. 117–192.
11. Hu, D. L., Chan, B., and Bush, J. W. M. (2003). The hydrodynamics of water strider locomotion, *Nature*, **424**, pp. 663–666.

References

12. Su, Y., Ji, B., Huang, Y., and Hwang, K.-C. (2010). Nature's design of hierarchical superhydrophobic surfaces of a water strider for low adhesion and low-energy dissipation, *Langmuir*, **26**, pp. 18926–18937.

13. Shirtcliffe, N. J., McHale, G., Newton, M. I., Perry, C. C., and Pyatt, F. B. (2006). Plastron properties of a superhydrophobic surface, *Appl. Phys. Lett.*, **89**, pp. 104106/1–104106/2.

14. Balmert, A., Bohn, H. F., Ditsche-Kuru, P., and Barthlott, W. (2011). Dry under water: Comparative morphology and functional aspects of air-retaining insect surfaces, *J. Morphol.*, **272**, pp. 442–451.

15. Feng, L., Zhang, Y., Xi, J., Zhu, Y., Wang, N., Xia, F., and Jiang, L. (2008). Petal effect: A superhydrophobic state with high adhesive force, *Langmuir*, **24**, pp. 4114–4119.

16. Bhushan, B., and Nosonovsky, M. (2010). The rose petal effect and the modes of superhydrophobicity, *Phil. Trans. R. Soc. A*, **368**, pp. 4713–4728.

17. Szczepanski, C. R., Darmanin, T., and Guittard, F. (2016). Spontaneous, phase-separation induced surface roughness: A new method to design parahydrophobic polymer coatings with rose petal-like morphology, *ACS Appl. Mater. Interfaces*, **8**, pp. 3063–3071.

18. Marmur, A. (2012). Hydro- hygro- oleo- omni-phobic? Terminology of wettability classification, *Soft Matter*, **8**, pp. 6867–6870.

19. Yang, S., Ju, J., Qiu, Y., He, Y., Wang, H., Dou, S., Liu, K., and Jiang, L. (2014). Peanut leaf inspired multifunctional surfaces, *Small*, **10**, pp. 294–299.

20. Lu, X., Cai, H., Wu, Y., Teng, C., Jiang, C., Zhu, Y., and Jiang, L. (2015). Peach skin effect: A quasi-superhydrophobic state with high adhesive force, *Sci. Bull.*, **60**, pp. 453–459.

21. Liu, K., Du, J., Wu J., and Jiang, L. (2012). Superhydrophobic gecko feet with high adhesive forces towards water and their bio-inspired materials, *Nanoscale*, **4**, pp. 768–772.

22. Wisdom, K. M., Watson, J. A., Qu, X., Liu, F., Watson, G. S., and Chen, C.-H. (2013). Self-cleaning of superhydrophobic surfaces by self-propelled jumping condensate, *Proc. Natl. Acad. Sci. U. S. A.*, **110**, pp. 7992–7997.

23. Sun, M., Liang, A., Watson, G. S., Watson, J. A., Zheng, Y., Ju, J., and Jiang, L. (2012). Influence of cuticle nanostructuring on the wetting behaviour/states on cicada wings, *PLoS ONE*, **7**, p. e35056.

24. Popp, T. M. O., Addison, J. B., Jordan, J. S., Damle, V. G., Rykaczewski, K., Chang, S. L. Y., Stokes, G. Y., Edgerly, J. S., and Yarger, J. L. (2016). Surface and wetting properties of Embiopteran (Webspinner) nanofiber silk, *Langmuir*, **32**, pp. 4681–4687.
25. Shirtcliffe, N. J., McHale, G., and Newton, M. I. (2009). Learning from superhydrophobic plants: The use of hydrophilic areas on superhydrophobic surfaces for droplet control, *Langmuir*, **25**, pp. 14121–14128.
26. Miele, E., Girardo, S., and Pisignano, D. (2012). Strelitzia reginae leaf as a natural template for anisotropic wetting and superhydrophobicity, *Langmuir*, **28**, pp. 5312–5317.
27. Feng, L., Li, S., Li, Y., Li, H., Zhang, L., Zhang, J., Song, Y., Liu, B., Jiang, L., and Zhu, D. (2002). Super-hydrophobic surfaces: From natural to artificial, *Adv. Mater.*, **14**, pp. 1857–1860.
28. Parker, A. R., and Lawrence, C. R. (2001). Water capture by a desert beetle, *Nature*, **414**, pp. 33–34.
29. Sun, J., and Bhushan, B. (2012). Structure and mechanical properties of beetle wings: A review, *RSC Adv.*, **2**, pp. 12606–12623.
30. Zhai, L., Berg, M. C., Cebeci, F. C., Kim, Y., Milwid, J. M., Rubner, M. F., and Cohen, R. E. (2006). Patterned superhydrophobic surfaces: Toward a synthetic mimic of the Namib desert beetle, *Nano Lett.*, **6**, pp. 1213–1217.
31. Zheng, Y., Gao, X., and Jiang, L. (2007). Directional adhesion of superhydrophobic butterfly wings, *Soft Matter*, **3**, pp. 178–182.
32. Kusumaatmaja H., and Yeomans, J. M. (2009). Anisotropic hysteresis on ratcheted superhydrophobic surfaces, *Soft Matter*, **5**, pp. 2704–2707.
33. Liu, C., Xue, Y., Chen, Y., and Zheng, Y. (2015). Effective directional self-gathering of drops on spine of cactus with splayed capillary arrays, *Sci. Rep.*, **5**, p. 17757.
34. Ju, J., Bai, H., Zheng, Y., Zhao, T., Fang, R., and Jiang, L. (2012). A multi-structural and multi-functional integrated fog collection system in cactus, *Nat. Commun.*, **3**, p. 1247.
35. Cao, M., Ju, J., Li, K., Dou, S., Liu, K., and Jiang, L. (2014). Facile and large-scale fabrication of a cactus-inspired continuous fog collector, *Adv. Funct. Mater.*, **24**, pp. 3235–3240.
36. Heng, X., Xiang, M., Lu, Z., and Luo, C. (2014). Branched ZnO wire structures for water collection inspired by cacti, *ACS Appl. Mater. Interfaces*, **6**, pp. 8032–8041.

37. Roth-Nebelsick, A., Ebner, M., Miranda, T., Gottschalk, V., Voigt, D., Gorb, S., Stegmaier, T., Sarsour, J., Linke, M., and Konrad, W. (2012). Leaf surface structures enable the endemic Namib desert grass Stipagrostis sabulicola to irrigate itself with fog water, *J. R. Soc. Interface*, **9**, pp. 1965–1974.
38. Andrews, H. G., Eccles, E. A., Schofield, W. C. E., and Badyal, J. P. S. (2011). Three-dimensional hierarchical structures for fog harvesting, *Langmuir*, **27**, pp. 3798–3802.
39. Hensel, R., Neinhuis, C., and Werner, C. (2016). The springtail cuticle as a blueprint for omniphobic surfaces, *Chem. Soc. Rev.*, **45**, pp. 323–341.
40. Nickerl, J., Tsurkan, M., Hensel, R., Neinhuis, C., and Werner, C. (2014). The multi-layered protective cuticle of Collembola: A chemical analysis, *J. R. Soc. Interface*, **11**, p. 20140619.
41. Nickerl, J., Helbig, R., Schulz, H.-J., Werner, C., and Neinhuis, C. (2013). Diversity and potential correlations to the function of Collembola cuticle structures, *Zoomorphology*, **132**, pp. 183–195.
42. Hensel, R., Helbig, R., Aland, S., Braun, H.-G., Voigt, A., Neinhuis, C., and Werner, C. (2011). Smart skin patterns protect springtails, *PLoS ONE*, **6**, e25105.
43. Hensel, R., Helbig, R., Aland, S., Braun, H.-G., Voigt, A., Neinhuis, C., and Werner, C. (2013). Wetting resistance at its topographical limit: The benefit of mushroom and serif T structures, *Langmuir*, **29**, pp. 1100–1112.
44. Gundersen, H., Leinaas, H. P., and Thaulow, C. (2014). Surface structure and wetting characteristics of Collembola cuticles, *PLoS ONE*, **9**, p. e102961.
45. Young, T. (1805). An essay on the cohesion of fluids, *Phil. Trans. R. Soc. Lond.*, **95**, pp. 65–87.
46. Wenzel, R. N. (1936). Resistance of solid surfaces to wetting by water, *Ind. Eng. Chem.*, **28**, pp. 988–994.
47. Cassie, A. B. D., and Baxter, S. (1944). Wettability of porous surfaces, *Trans. Faraday Soc.*, **40**, pp. 546–551.
48. Papadopoulou, S. K., Tsioptsias, C., Pavlou, A., Kaderides, K., Sotiriou, S., and Panayiotou, C. (2011). Superhydrophobic surfaces from hydrophobic or hydrophilic polymers via nanophase separation or electrospinning/electrospraying, *Colloids Surf. A*, **387**, pp. 71–78.
49. Su, Y., Ji, B., Zhang, K., Gao, H., Huang, Y., and Hwang, K. (2010). Nano to micro structural hierarchy is crucial for stable

superhydrophobic and water-repellent surfaces, *Langmuir*, **26**, pp. 4984–4989.
50. Yu, Y., Zhao, Z.-H., and Zheng, Q.-S. (2007). Mechanical and superhydrophobic stabilities of two-scale surfacial structure of lotus leaves, *Langmuir*, **23**, pp. 8212–8216.
51. Sarkar, A., and Kietzig, A.-M. (2015). Design of a robust superhydrophobic surface: Thermodynamic and kinetic analysis, *Soft Matter*, **11**, pp. 1998–2007.
52. Watson, G. S., Cribb, B. W., and Watson, J. A. (2010). How micro/nanoarchitecture facilitates anti-wetting: An elegant hierarchical design on the termite wing, *ACS Nano*, **4**, pp. 129–136.
53. Bellanger, H., Darmanin, T., Taffin de Givenchy, E., and Guittard, F. (2014). Chemical and physical pathways for the preparation of superoleophobic surfaces and related wetting theories, *Chem. Rev.*, **114**, pp. 2694–2716.
54. Liu, T., and Kim, C.-J. (2014). Turning a surface superrepellent even to completely wetting liquids, *Science*, **346**, pp. 1096–1100.
55. Tuteja, A., Choi, W., Ma, M., Mabry, J. M., Mazzella, S. A., Rutledge, G. C., McKinley, G. H., and Cohen, R. E. (2007). Designing superoleophobic surfaces, *Science*, **318**, pp. 1618–1622.
56. Darmanin, T., and Guittard, F. (2009). Molecular design of conductive polymers to modulate superoleophobic properties, *J. Am. Chem. Soc.*, **131**, pp. 7928–793.
57. Darmanin, T., and Guittard, F. (2014). Enhancement of the superoleophobic properties of fluorinated PEDOP using polar glycol spacers, *J. Phys. Chem. C*, **118**, pp. 26912–26920.
58. Yin, B., Fang, L., Hu, J., Tang, A. Q., Hea, J., and Mao, J. H. (2012). A facile method for fabrication of superhydrophobic coating on aluminum alloy, *Surf. Interface Anal.*, **44**, pp. 439–444.
59. Liao, R., Zuo, Z., Guo, C., Yuan, Y., and Zhuang, A. (2014). Fabrication of superhydrophobic surface on aluminum by continuous chemical etching and its anti-icing property, *Appl. Surf. Sci.*, **317**, pp. 701–709.
60. Esmaeilirad, A., Rukosuyev, M. V., Jun, M. B. G., and van Veggel, F. C. J. M. (2016). A cost-effective method to create physically and thermally stable and storable super-hydrophobic aluminum alloy surfaces, *Surf. Coat. Technol.*, **285**, pp. 227–234.
61. Peng, S., and Bhushan, B. (2016). Mechanically durable superoleophobic aluminum surfaces with microstep and nanoreticula hierarchical structure for self-cleaning and anti-smudge properties, *J. Colloid Interface Sci.*, **461**, pp. 273–284.

62. Li, X., Zhang, Q., Guo, Z., Shi, T., Yu, J., Tang, M., and Huang, X. (2015). Fabrication of superhydrophobic surface with improved corrosion inhibition on 6061 aluminum alloy substrate, *Appl. Surf. Sci.*, **342**, pp. 76–83.
63. Li, X.-W., Zhang, Q.-X., Guo, Z., Yu, J.-G., Tang, M.-K., and Huang, X.-J. (2015). Low-cost and large-scale fabrication of a superhydrophobic 5052 aluminum alloy surface with enhanced corrosion resistance, *RSC Adv.*, **5**, pp. 29639–29646.
64. Wang, Y., Liu, X. W., Zhang, H. F., and Zhou, Z. P. (2015). Fabrication of self-healing super-hydrophobic surfaces on aluminium alloy substrates, *AIP Adv.*, **5**, pp. 041314/1–041314/6.
65. Zhang, Y., Wu, J., Yu, X., and Wu, H. (2011). Low-cost one-step fabrication of superhydrophobic surface on Al alloy, *Appl. Surf. Sci.*, **257**, pp. 7928–7931.
66. Rezayi, T., and Entezari, M. H. (2016). Toward a durable superhydrophobic aluminum surface by etching and ZnO nanoparticle deposition, *J. Colloid Interface Sci.*, **463**, pp. 37–45.
67. Chu, F., and Wu, X. (2016). Fabrication and condensation characteristics of metallic superhydrophobic surface with hierarchical micro-nano structures, *Appl. Surf. Sci.*, **371**, pp. 322–328.
68. Kwak, W., and Hwang, W. (2016). Facile method for preparing superoleophobic surfaces with hierarchical microcubic/nanowire structures, *Nanotechnology*, **27**, pp. 055301/1–055301/8.
69. Frankiewicz, C., and Attinger, D. (2016). Texture and wettability of metallic lotus leaves, *Nanoscale*, **8**, pp. 3982–3990.
70. Guo, J., Yang, F., and Guo, Z. (2016). Fabrication of stable and durable superhydrophobic surface on copper substrates for oil–water separation and ice-over delay, *J. Colloid Interface Sci.*, **466**, pp. 36–43.
71. Jie, H., Xu, Q., Wei, L., and Min, Y. (2016). Etching and heating treatment combined approach for superhydrophobic surface on brass substrates and the consequent corrosion resistance, *Corros. Sci.*, **102**, pp. 251–258.
72. Liu, T., Chen, S., Cheng, S., Tian, J., Chang, X., and Yin, Y. (2007). Corrosion behavior of super-hydrophobic surface on copper in seawater, *Electrochim. Acta*, **52**, pp. 8003–8007.
73. Liu, T., Yin, Y., Chen, S., Chang, X., and Cheng, S. (2007). Superhydrophobic surfaces improve corrosion resistance of copper in seawater, *Electrochim. Acta*, **52**, pp. 3709–3713.

74. Wang, S., Feng, L., and Jiang, L. (2006). One-step solution-immersion process for the fabrication of stable sionic superhydrophobic surfaces, *Adv. Mater.*, **18**, pp. 767–7770.
75. Chen, Y., Chen, S., Yu, F., Sun, W., Zhu, H., and Yin, Y. (2009). Fabrication and anti-corrosion property of superhydrophobic hybrid film on copper surface and its formation mechanism, *Surf. Interface Anal.*, **41**, pp. 872–877.
76. Chen, X. H., Yang, G. B., Kong, L. H., Dong, D., Yu, L. G., Chen, J. M., and Zhang, P. Y. (2009). Direct growth of hydroxy cupric phosphate heptahydrate monocrystal with honeycomb-like porous structures on copper surface mimicking Lotus leaf, *Cryst. Growth Des.*, **9**, pp. 2656–2661.
77. Liang, M., Wei, Y., Hou, L., Wang, H., Li, Y., and Guo, C. (2016). Fabrication of a super-hydrophobic surface on a magnesium alloy by a simple method, *J. Alloys Compd.*, **656**, pp. 311–317.
78. Wang, Y., Wang, W., Zhong, L., Wang, J., Jiang, Q., and Guo, X. (2010). Super-hydrophobic surface on pure magnesium substrate by wet chemical method, *Appl. Surf. Sci.*, **256**, pp. 3837–3840.
79. Feng, L., Zhu, Y., Fan, W., Wang, Y., Qiang, X., and Liu Y. (2015). Fabrication and corrosion resistance of superhydrophobic magnesium alloy, *Appl. Phys. A*, **120**, pp. 561–570.
80. Kim, Y., Go, S., and Ahn Y. (2013). Fabrication of a superhydrophobic surface with flower-like microstructures with a one-step immersion process, *Bull. Korean Chem. Soc.*, **34**, pp. 3495–3498.
81. Wan, H., and Hu, X. (2016). One-step solve-thermal process for the construction of anticorrosion bionic superhydrophobic surfaces on magnesium alloy, *Mater. Lett.*, **174**, pp. 209–212.
82. Wang, H., Yang, Z., Yu, J., Wu, Y., Shao, W., Jiang, T., and Xu, X. (2014). Preparation of lotus-like hierarchical microstructures on zinc substrate and study of its wettability, *RSC Adv.*, **4**, pp. 33730–33738.
83. Ren, G., Zhang, Z., Zhu, X., Ge, B., Wang, K., Xu, X., Men, X., and Zhou, X. (2014). A facile method for imparting superoleophobicity to polymer substrates, *Appl. Phys. A*, **114**, pp. 1129–1133.
84. Meng, H., Wang, S., Xi, J., Tang, Z., and Jiang, L. (2008). Facile means of preparing superamphiphobic surfaces on common engineering metals, *J. Phys. Chem. C*, **112**, pp. 11454–11458.
85. Liu, H., Szunerits, S., Xu, W., and Boukherroub, R. (2009). Preparation of superhydrophobic coatings on zinc as effective corrosion barriers, *ACS Appl. Mater. Interfaces*, **1**, pp. 1150–1153.

86. Liu, H., Szunerits, S., Pisarek, M., Xu, W., and Boukherroub, R. (2009). Preparation of superhydrophobic coatings on zinc, silicon, and steel by a solution-immersion technique, *ACS Appl. Mater. Interfaces*, **1**, pp. 2086–2091.
87. Yu, S. R., Liu, J. A., Diao, W., and Li, W. (2014). Preparation of a bionic microtexture on X52 pipeline steels and its superhydrophobic behavior, *J. Alloys Compd.*, **585**, pp. 689–695.
88. Park, B., and Hwang, W. (2016). A facile fabrication method for corrosion-resistant micro/nanostructures on stainless steel surfaces with tunable wettability, *Scr. Mater.*, **113**, pp. 118–121.
89. Zhang, H., Yang, J., Chen, B., Liu, C., Zhang, M., and Li, C. (2015). Fabrication of superhydrophobic textured steel surface for anti-corrosion and tribological properties, *Appl. Surf. Sci.*, **359**, pp. 905–910.
90. Lu, Y., Sathasivam, S., Song, J., Chen, F., Xu, W., Carmalt, C. J., and Parkin, I. P. (2014). Creating superhydrophobic mild steel surfaces for water proofing and oil–water separation, *J. Mater. Chem. A*, **2**, pp. 11628–11634.
91. Qu, M., Zhang, B., Song, S., Chen, L., Zhang, J., and Cao, X. (2007). Fabrication of superhydrophobic surfaces on engineering materials by a solution-immersion process, *Adv. Funct. Mater.*, **17**, pp. 593–596.
92. Grynyov, R., Bormashenko, E., Whyman, G., Bormashenko, Y., Musin, A., Pogreb, R., Starostin, A., Vatsifer, V., Strelnikov, V., Schechter, A., and Kolagatla, S. (2016). Superoleophobic surfaces obtained via hierarchical metallic meshes, *Langmuir*, **32**, pp. 4134–4140.
93. Liu, Y., Zhang, K., Yao, W., Liu, J., Han, Z., and Ren, L. (2016). Bioinspired structured superhydrophobic and superoleophilic stainless steel mesh for efficient oil-water separation, *Colloids Surf. A*, **500**, pp. 54–63.
94. Song, J.-l., Xu, W.-J., Liu, X., Lu, Y., and Sun, J. (2012). Electrochemical machining of super-hydrophobic Al surfaces and effect of processing parameters on wettability, *Appl. Phys. A*, **108**, pp. 559–568.
95. Lu, Y., Xu, W., Song, J., Liu, X., Xing, Y., and Sun, J. (2012). Preparation of superhydrophobic titanium surfaces via electrochemical etching and fluorosilane modification, *Appl. Surf. Sci.*, **263**, pp. 297–301.
96. Sun, J., Zhang, F., Song, J., Wang, L., Qu, Q., Lu, Y., and Parkin, I. (2014). Electrochemical fabrication of superhydrophobic Zn surfaces, *Appl. Surf. Sci.*, **315**, pp. 346–352.
97. Yu, H., Lian, Z., Wan, Y., Weng, Z., Xu, J., and Yu, Z. (2015). Fabrication of durable superamphiphobic aluminum alloy surfaces with

anisotropic sliding by HS-WEDM and solution immersion processes, *Surf. Coat. Technol.*, **275**, pp. 112–119.

98. Wang, L., Yang, J., Zhu, Y., Li, Z., Shen, T., and Yang, D.-Q. (2016). An environment-friendly fabrication of superhydrophobic surfaces on steel and magnesium alloy, *Mater. Lett.*, **171**, pp. 297–299.

99. Wang, F., Zhao, K., Cheng, J., and Zhang, J. (2011). Conciliating surface superhydrophobicities and mechanical strength of porous silicon films, *Appl. Surf. Sci.*, **257**, pp. 2752–2755.

100. Nenzi, P., Giacomello, A., Bolognesi, G., Chinappi, M., Balucani, M., and Casciola, C. M. (2011). Superhydrophobic porous silicon surfaces, *Sens. Transducers J.*, **13**, pp. 62–72.

101. Wang, M.-F., Raghunathan, N., Ziaie, B. (2007). A nonlithographic top-down electrochemical approach for creating hierarchical (micro–nano) superhydrophobic silicon surfaces, *Langmuir*, **23**, pp. 2300–2303.

102. Seo, J., Lee, S., Han, H., Chung, Y., Lee, J., Kim, S.-D., Kim, Y.-W., Lim, S., and Lee, T. (2013). Reversible wettability control of silicon nanowire surfaces: From superhydrophilicity to superhydrophobicity, *Thin Solid Films*, **527**, pp. 179–185.

103. Wang, R.-C., Chao, C.-Y., and Su W.-S. (2012). Electrochemically controlled fabrication of lightly doped porous Si nanowire arrays with excellent antireflective and self-cleaning properties, *Acta Mater.*, **60**, pp. 2097–2103.

104. Kim, B. S., Shin, S., Shin, S. J., Kim, K. M., and Cho, H. H. (2011). Control of superhydrophilicity/superhydrophobicity using silicon nanowires via electroless etching method and fluorine carbon coatings, *Langmuir*, **27**, pp. 10148–10156.

105. Egatz-Gomez, A., Majithia, R., Levert, C., and Meissner, K. E. (2012). Super-wetting, wafer-sized silicon nanowire surfaces with hierarchical roughness and low defects, *RSC Adv.*, **2**, pp. 11472–11480.

106. Yoon, S.-S., and Khang, D.-Y. (2012). Switchable wettability of vertical Si nanowire array surface by simple contact-printing of siloxane oligomers and chemical washing, *J. Mater. Chem.*, **22**, pp. 10625–10630.

107. Chen, C.-Y., and Wong, C.-P. (2015). Unveiling the shape-diversified silicon nanowires made by HF/HNO$_3$ isotropic etching with the assistance of silver, *Nanoscale*, **7**, pp. 1216–1223.

108. Nguyen, T. P. N., Dufour, R., Thomy, V., Senez, V., Boukherroub, R., and Coffinier, Y. (2014). Fabrication of superhydrophobic and highly oleophobic silicon-based surfaces via electroless etching method, *Appl. Surf. Sci.*, **295**, pp. 38–43.

109. Dawood, M. K., Zheng, H., Liew, T. H., Leong, K. C., Foo, Y. L., Rajagopalan, R., Khan, S. A., and Choi, W. K. (2011). Mimicking both Petal and Lotus effects on a single silicon substrate by tuning the wettability of nanostructured surfaces, *Langmuir*, **27**, pp. 4126–4133.

110. Dawood, M. K., Zheng, H., Kurniawan, N. A., Leong, K. C., Foo, Y. L., Rajagopalan, R., Khanbe, S. A., and Choi, W. K. (2012). Modulation of surface wettability of superhydrophobic substrates using Si nanowire arrays and capillary-force-induced nanocohesion, *Soft Matter*, **8**, pp. 3549–3557.

111. Piret, G., Coffinier, Y., Roux, C., Melnyk, O., and Boukherroub, R. (2008). Biomolecule and nanoparticle transfer on patterned and heterogeneously wetted superhydrophobic silicon nanowire surfaces, *Langmuir*, **24**, pp. 1670–1672.

112. Kuan, W.-F., and Chen, L.-J. (2009). The preparation of super-hydrophobic surfaces of hierarchical silicon nanowire structures, *Nanotechnology*, **20**, pp. 035605/1–035605/8.

113. Lo, C.-W., Wang, C.-C., and Lu, M.-C. (2014). Spatial control of heterogeneous nucleation on the superhydrophobic nanowire array, *Adv. Funct. Mater.*, **24**, pp. 1211–1217.

114. Seo, J., Lee, S., Han, H., Jung, H. B., Hong, J., Song, G., Cho, S. M., Park, C., Lee, W., and Lee, T. (2013). Gas-driven ultrafast reversible switching of super-hydrophobic adhesion on palladium-coated silicon nanowires, *Adv. Mater.*, **25**, pp. 4139–4144.

115. Seo, J., Lee, J. S., Lee, K., Kim, D., Yang, K., Shin, S., Mahata, C., Jung, H. B., Lee, W., Cho, S.-W., and Lee, T. (2014). Switchable water-adhesive, superhydrophobic palladium-layered silicon nanowires potentiate the angiogenic efficacy of human stem cell spheroids, *Adv. Mater.*, **26**, pp. 7043–7050.

116. Xu, C., Song, Z., Jin, J., Xiang, Q., and Feng, X. (2016). A high performance three-phase enzyme electrode based on superhydrophobic mesoporous silicon nanowire arrays for glucose detection, *Nanoscale*, **8**, pp. 7391–7395.

117. Gentile, F., Coluccio, M. L., Zaccaria, R. P., Francardi, M., Cojoc, G., Perozziello, G., Raimondo, R., Candeloro, P., and Di Fabrizio, E. (2014). Selective on site separation and detection of molecules in diluted solutions with super-hydrophobic clusters of plasmonic nanoparticles, *Nanoscale*, **6**, pp. 8208–8225.

118. Ko, T.-J., Her, E. K., Shin, B., Kim, H.-Y., Lee, K.-R., Hong, B. K., Kim, S. H., Oh, K. H., and Moon, M.-W. (2012). Water condensation behavior

on the surface of a network of superhydrophobic carbon fibers with high-aspect-ratio nanostructures, *Carbon*, **50**, pp. 5085–5092.
119. Kumar, R. T. R., Mogensen, K. B., and Bøggild, P. (2010). Simple approach to superamphiphobic overhanging silicon nanostructures, *J. Phys. Chem. C*, **114**, pp. 2936–2940.
120. Rajkumar, K., and Rajendrakumar, R. T. (2013). Fabrication and Electrowetting Properties of Poly Si Nanostructure Based Superhydrophobic Platform, *Plasma Chem. Plasma Process.*, **33**, pp. 807–816.
121. Zhang, X.-S., Meng, B., Zhu, F.-Y., Tang, W., and Zhang, H.-X. (2014). Switchable wetting and flexible SiC thin film with nanostructures for microfluidic surface-enhanced Raman scattering sensors, *Sens. Actuators A*, **208**, pp. 166–173.
122. Liu, Y., Lin, W., Lin, Z., Xiu, Y., and Wong, C. P. (2012). A combined etching process toward robust superhydrophobic SiC surfaces, *Nanotechnology*, **23**, pp. 255703/1–255703/7.
123. Marcon, L., Addad, A., Coffinier, Y., and Boukherroub, R. (2013). Cell micropatterning on superhydrophobic diamond nanowires, *Acta Biomater.*, **9**, pp. 4585–4591.
124. Zhou, Y. B., Yang, Y., Liu, W. M., Ye, Q., He, B., Zou, Y. S., Wang, P. F., Pan, X. J., Zhang, W. J., Bello, I., and Lee, S. T. (2010). Preparation of superhydrophobic nanodiamond and cubic boron nitride films, *Appl. Phys. Lett.*, **97**, pp. 133110/1–133110/3.
125. Kondrashov, V., and Ruehe, J. (2014). Microcones and nanograss: Toward mechanically robust superhydrophobic surfaces, *Langmuir*, **30**, pp. 4342–4350.
126. Park, Y.-B., Im, M., Im, H., and Choi, Y.-K. (2010). Superhydrophobic cylindrical nanoshell array, *Langmuir*, **26**, pp. 7661–7664.
127. Zhilei, C., Maobing, S., and Lida, W. (2013). Cathodic etching for fabrication of super-hydrophobic aluminum coating with micro/nano-hierarchical structure, *J. Solid State Electrochem.*, **17**, pp. 2661–2669.
128. Pei, M., Wang, B., Tang, Y., Song, X., Yan, H., and Zhang, X. (2013). Fabrication of superhydrophobic copper surface by direct current sputtering and its underwater stability, *Thin Solid Films*, **548**, pp. 313–316.
129. Liao, R., Zuo, Z., Guo, C., Zhuang, A., Zhao, X., and Yuan, Y. (2015). Anti-icing performance in glaze ice of nanostructured film prepared by RF magnetron sputtering, *Appl. Surf. Sci.*, **356**, pp. 539–545.

130. Zhou, X., Xu, D., Lu, J., and Zhang, K. (2015). CuO/Mg/fluorocarbon sandwich-structure superhydrophobic nanoenergetic composite with anti-humidity property, *Chem. Eng. J.*, **266**, pp. 163–170.
131. Wang, B., and Guo, Z. (2013). pH-responsive bidirectional oil–water separation material, *Chem. Commun.*, **49**, pp. 9416–9418.
132. Macias-Montero, M., Borras, A., Romero-Gomez, P., Cotrino, J., Frutos, F., and Gonzalez-Elipe, A. R. (2014). Plasma deposition of superhydrophobic Ag@TiO$_2$ core@shell nanorods on processable substrates, *Plasma Process. Polym.*, **11**, pp. 164–174.
133. Kamegawa, T., Shimizu, Y., and Yamashita, H. (2012). Superhydrophobic surfaces with photocatalytic self-cleaning properties by nanocomposite coating of TiO$_2$ and polytetrafluoroethylene, *Adv. Mater.*, **24**, pp. 3697–3700.
134. Aytug, T., Bogorin, D. F., Paranthaman, P. M., Mathis, J. E., Simpson, J. T., and Christen, D. K. (2014). Superhydrophobic ceramic coatings enabled by phase-separated nanostructured composite TiO$_2$–Cu$_2$O thin films, *Nanotechnology*, **25**, pp. 245601/1–245601/7.
135. Singh, D. P., and Singh, J. P. (2014). Controlled growth of standing Ag nanorod arrays on bare Si substrate using glancing angle deposition for self-cleaning applications, *Appl. Phys. A*, **114**, pp. 1189–1193.
136. Kumar, S., Goel, P., Singh, D. P., and Singh, J. P. (2014). Highly sensitive superhydrophobic Ag nanorods array substrates for surface enhanced fluorescence studies, *Appl. Phys. Lett.*, **104**, pp. 023107/1–023107/4.
137. Goel, P., Kumar, S., Sarkar, J., and Singh, J. P. (2015). Mechanical strain induced tunable anisotropic wetting on buckled PDMS silver nanorods arrays, *ACS Appl. Mater. Interfaces*, **7**, pp. 8419–8426.
138. Hall, J. Z., Taschuk, M. T., and Brett, M. J. (2012). Polarity-adjustable reversed phase ultrathin-layer chromatography, *J. Chromatogr. A*, **1266**, pp. 168–174.
139. Zhou, X., Xu, D., Yang, G., Zhang, Q., Shen, J., Lu, J., and Zhang, K. (2014). Highly exothermic and superhydrophobic Mg/fluorocarbon core/shell nanoenergetic arrays, *ACS Appl. Mater. Interfaces*, **6**, pp. 10497–10505.
140. Gwon, H. J., Park, Y., Moon, C. W., Nahm, S., Yoon, S.-J., Kim, S. Y., and Jang, H. W. (2014). Superhydrophobic and antireflective nanograss-coated glass for high performance solar cells, *Nano Res.*, **7**, pp. 670–678.
141. Kuang, P., Hsieh, M.-L., and Lin, S.-Y. (2015). Integrated three-dimensional photonic nanostructures for achieving near-unity

solar absorption and superhydrophobicity, *J. Appl. Phys.*, **117**, pp. 215309/1–215309/6.

142. Tsoi, S., Fok, E., Sit, J. C., and Veinot, J. G. C. (2006). Surface functionalization of porous nanostructured metal oxide thin films fabricated by glancing angle deposition, *Chem. Mater.*, **18**, pp. 5260–5266.

143. Tsoi, S., Fok, E., Sit, J. C., and Veinot, J. G. C. (2004). Superhydrophobic, high surface area, 3-D SiO_2 nanostructures through siloxane-based surface functionalization, *Langmuir*, **20**, pp. 10771–10774.

144. Khedir, K. R., Kannarpady, G. K., Ishihara, H., Woo, J., Ryerson, C., and Biris, A. S. (2011). Design and fabrication of Teflon-coated tungsten nanorods for tunable hydrophobicity, *Langmuir*, **27**, pp. 4661–4668.

145. Kannarpady, G. K., Khedir, K. R., Ishihara, H., Woo, J., Oshin, O. D., Trigwell, S., Ryerson, C., and Biris, A. S. (2011). Controlled growth of self-organized hexagonal arrays of metallic nanorods using template-assisted glancing angle deposition for superhydrophobic applications, *ACS Appl. Mater. Interfaces*, **3**, pp. 2332–2340.

146. Khedir, K. R., Kannarpady, G. K., Ishihara, H., Woo, J., Ryerson, C., and Biris, A. S. (2010). Morphology control of tungsten nanorods grown by glancing angle RF magnetron sputtering under variable argon pressure and flow rate, *Phys. Lett. A*, **374**, pp. 4430–4437.

147. Khedir, K. R., Kannarpady, G. K., Ishihara, H., Woo, J., Trigwell, S., Ryerson, C., and Biris, A. S. (2011). Advanced studies of water evaporation kinetics over Teflon-coated tungsten nanorod surfaces with variable hydrophobicity and morphology, *J. Phys. Chem. C*, **115**, pp. 13804–13812.

148. Xiao, X., Cao, G., Chen, F., Tang, Y., Liu, X., and Xu, W. (2015). Durable superhydrophobic wood fabrics coating with nanoscale Al_2O_3 layer by atomic layer deposition, *Appl. Surf. Sci.*, **349**, pp. 876–879.

149. Ding, Y., Xu, S., Zhang, Y., Wang, A. C., Wang, M. H., Xiu, Y., Wong, C. P., and Wang, Z. L. (2008). Modifying the anti-wetting properties of butterfly wings and water strider legs by atomic layer deposition coating: Surface materials versus geometry, *Nanotechnology*, **19**, pp. 355708/1–355708/7.

150. Szilagyi, I. M., Teucher, G., Harkonen, E., Farm, E., Hatanpaa, T., Nikitin, T., Khriachtchev, L., Rasanen, M., Ritala, M., and Leskela, M. (2013). Programming nanostructured soft biological surfaces by atomic layer deposition, *Nanotechnology*, **24**, pp. 245701/1–245701/6.

151. Cao, Y., Deng, S., Hu, Q., Zhong, Q., Luo, Q.-P., Yuan, L., and Zhou, J. (2015). Three-dimensional ZnO porous films for self-cleaning ultraviolet photodetectors, *RSC Adv.*, **5**, pp. 85969–85973.
152. Oh, I.-K., Kim, K., Lee, Z., Ko, K. Y., Lee, C.-W., Lee, S. J., Myung, J. M., Lansalot-Matras C., Noh, W., Dussarat, C., Kim, H., and Lee, H.-B.-R. (2015). Hydrophobicity of rare earth oxides grown by atomic layer deposition, *Chem. Mater.*, **27**, pp. 148–156.
153. Datta, D. P., and Som, T. (2016). Nanoporosity-induced superhydrophobic and large antireflection in InSb, *Appl. Phys. Lett.*, **108**, pp. 191603/1–191603/5.
154. Datta, D. P., Garg, S. K., Thakur, I., Satpati, B., Sahoo, P. K., Kanjilal, D., and Som, T. (2016). Facile synthesis of a superhydrophobic and colossal broadband antireflective nanoporous GaSb surface, *RSC Adv.*, **6**, pp. 48919–48926.
155. Yong, J., Chen, F., Yang, Q., and Hou, X. (2015). Femtosecond lase controlled wettability of solid surfaces, *Soft Matter*, **11**, pp. 8897–8906.
156. Chen, F., Zhang, D., Yang, Q., Yong, J., Du, G., Si, J., Yun, F., and Hou, X. (2013). Bioinspired wetting surface via laser microfabrication, *ACS Appl. Mater. Interfaces*, **5**, pp. 6777–6792.
157. Li, B.-J., Zhou, M., Zhang, W., Amoako, G., and Gao, C.-Y. (2012). Comparison of structures and hydrophobicity of femtosecond and nanosecond laser-etched surfaces on silicon, *Appl. Surf. Sci.*, **263**, pp. 45–49.
158. Ta, D. V., Dunn, A., Wasley, T. J., Kay, R. W., Stringer, J., Smith, P. J., Connaughton, C., and Shephard, J. D. (2015). Nanosecond laser textured superhydrophobic metallic surfaces and their chemical sensing applications, *Appl. Surf. Sci.*, **357**, pp. 248–254.
159. Boinovich, L. B., Emelyanenko, A. M., Modestov, A. D., Domantovsky, A. G., and Emelyanenko, K. A. (2015). Synergistic effect of superhydrophobicity and oxidized layers on corrosion resistance of aluminum alloy surface textured by nanosecond laser treatment, *ACS Appl. Mater. Interfaces*, **7**, pp. 19500–19508.
160. Boinovich, L. B., Domantovskiy, A. G., Emelyanenko, A. M., Pashinin, A. S., Ionin, A. A., Kudryashov, S. I., and Saltuganov, P. N. (2014). Femtosecond laser treatment for the design of electro-insulating superhydrophobic coatings with enhanced wear resistance on glass, *ACS Appl. Mater. Interfaces*, **6**, pp. 2080–2085.
161. Rukosuyev, M. V., Lee, J., Cho, S. J., Lim, G., and Jun, M. B. G. (2014). One-step fabrication of superhydrophobic hierarchical structures by femtosecond laser ablation, *Appl. Surf. Sci.*, **313**, pp. 411–417.

162. Emelyanenko, A. M., Shagieva, F. M., Domantovsky, A. G., and Boinovich, L. B. (2015). Nanosecond laser micro- and nanotexturing for the design of a superhydrophobic coating robust against long-term contact with water, cavitation, and abrasion, *Appl. Surf. Sci.*, **332**, pp. 513–517.

163. Vorobyev, A. Y., and Guo, C. (2015). Multifunctional surfaces produced by femtosecond laser pulses, *J. Appl. Phys.*, **117**, pp. 033103/1–033103/5.

164. Azimi, G., Kwon, H.-M., and Varanasi, K. K. (2014). Superhydrophobic surfaces by laser ablation of rare-earth oxide ceramics, *MRS Commun.*, 4, pp. 95–99.

165. Wang, J.-N., Shao, R.-Q., Zhang, Y.-L., Guo, L., Jiang, H.-B., Lu, D.-X., and Sun, H.-B. (2012). Biomimetic graphene surfaces with superhydrophobicity and iridescence, *Chem. Asian J.*, **7**, pp. 301–304.

166. Wang, J.-N., Shao, R.-Q., Zhang, Y.-L., Guo, L., Jiang, H.-B., Lu, D.-X., and Sun, H.-B. (2015). Superhydrophobic SERS substrates based on silver-coated reduced graphene oxide gratings prepared by two-beam laser interference, *ACS Appl. Mater. Interfaces*, **7**, pp. 27059–27065.

167. Pan, H., Luo, F., Lin, G., Wang, C., Dong, M., Liao, Y., and Zhao, Q.-Z. (2015). Quasi-superhydrophobic porous silicon surface fabricated by ultrashort pulsed-laser ablation and chemical etching, *Chem. Phys. Lett.*, **637**, pp. 159–163.

168. Jiang, H.-B., Zhang, Y.-L., Liu, Y., Fu, X.-Y., Li, Y.-F., Liu, Y.-Q., Li, C.-H., and Sun, H.-B. (2016). Bioinspired few-layer graphene prepared by chemical vapor deposition on femtosecond laser-structured Cu foil, *Laser Photonics Rev.*, **10**, pp. 441–450.

169. Li, H., Lai, Y., Huang, J., Tang, Y., Yang, L., Chen, Z., Zhang, K., Wang, X., and Tan, L. P. (2015). Multifunctional wettability patterns prepared by laser processing on superhydrophobic TiO_2 nanostructured surfaces, *J. Mater. Chem. B*, **3**, pp. 342–347.

170. Jiang, W., He, X., Liu, H., Yin, L., Shi, Y., and Ding, Y. (2014). Digital selective fabrication of micro/nano-composite structured TiO_2 nanorod arrays by laser direct writing, *J. Micromech. Microeng.*, **24**, pp. 115005/1–115005/11.

171. Long, J., Fan, P., Gong, D., Jiang, D., Zhang, H., Li, L., and Zhong, M. (2015). Superhydrophobic surfaces fabricated by femtosecond laser with tunable water adhesion: from Lotus leaf to Rose petal, *ACS Appl. Mater. Interfaces*, **7**, pp. 9858–9865.

172. Long, J., Pan, L., Fan, P., Gong, D., Jiang, D., Zhang, H., Li, L., and Zhong, M. (2016). Cassie-state stability of metallic superhydrophobic surfaces with various micro/nanostructures produced by a femtosecond laser, *Langmuir*, **32**, pp. 1065–1072.
173. Dong, C., Gu, Y., Zhong, M., Li, L., Sezer, K., Ma, M., and Liu, W. (2011). Fabrication of superhydrophobic Cu surfaces with tunable regular micro and random nano-scale structures by hybrid laser texture and chemical etching, *J. Mater. Process. Technol.*, **211**, pp. 1234–1240.
174. Zhang, D., Chen, F., Yang, Q., Yong, J., Bian, H., Ou, Y., Si, J., Meng, X., and Hou, X. (2012). A simple way to achieve pattern-dependent tunable adhesion in superhydrophobic surfaces by a femtosecond laser, *ACS Appl. Mater. Interfaces*, **4**, pp. 4905–4912.
175. Qin, L., Lin, P., Zhang, Y., Dong, G., and Zeng, Q. (2013). Influence of surface wettability on the tribological properties of laser textured Co–Cr–Mo alloy in aqueous bovine serum albumin solution, *Appl. Surf. Sci.*, **268**, pp. 79–86.
176. Zhou, W., Liu, W., Liu, S., Zhang, G., and Shen, Z. (2016). Experimental investigation on surface wettability of copper-based dry bioelectrodes, *Sens. Actuators A*, **244**, pp. 237–242.
177. Ahmmed, K. M. T., and Kietzig, A.-M. (2016). Drag reduction on laser-patterned hierarchical superhydrophobic surfaces, *Soft Matter*, **12**, 4912–4922.
178. Yong, J., Chen, F., Yang, Q., Farooq, U., and Hou, X. (2015). Photoinduced switchable underwater superoleophobicity–superoleophilicity on laser modified titanium surfaces, *J. Mater. Chem. A*, **3**, pp. 10703–10709.
179. Chen, F., Zhang, D., Yang, Q., Wang, X., Dai, B., Li, X., Hao, X., Ding, Y., Si, J., and Hou, X. (2011). Anisotropic wetting on microstrips surface fabricated by femtosecond laser, *Langmuir*, **27**, pp. 359–365.
180. Yong, J., Yang, Q., Chen, F., Zhang, D., Bian, H., Ou, Y., Si, J., Du, G., and Hou, X. (2013). Stable superhydrophobic surface with hierarchical mesh-porous structure fabricated by a femtosecond laser, *Appl. Phys. A*, **111**, pp. 243–249.
181. Long, J., Zhong, M., Zhang, H., and Fan, P. (2015). Superhydrophilicity to superhydrophobicity transition of picosecond laser microstructured aluminum in ambient air, *J. Colloid Interface Sci.*, **441**, pp. 1–9.
182. Tang, M.-K., Huang, X.-J., Guo, Z., Yu, J.-G., Li, X.-W., and Zhang, Q.-X. (2015). Fabrication of robust and stable superhydrophobic surface by a convenient, low-cost and efficient laser marking approach, *Colloids Surf. A*, **484**, pp. 449–456.

183. Yong, J., Chen, F., Yang, Q., Fang, Y., Huo, J., and Hou, X. (2015). Femtosecond laser induced hierarchical ZnO superhydrophobic surfaces with switchable wettability, *Chem. Commun.*, **51**, pp. 9813–9816.

184. Ta, V. D., Dunn, A., Wasley, T. J., Lib, J., Kay, R. W., Stringer, J., Smith, P. J., Esenturk, E., Connaughton, C., and Shephard, J. D. (2016). Laser textured superhydrophobic surfaces and their applications for homogeneous spot deposition, *Appl. Surf. Sci.*, **365**, pp. 153–159.

185. Liu, Y., Liu, J., Li, S., Han, Z., Yu, S., and Ren, L. (2014). Fabrication of biomimetic super-hydrophobic surface on aluminum alloy, *J. Mater. Sci.*, **49**, pp. 1624–1629.

186. Jagdheesh, R., García-Ballesteros, J. J., and Ocaña, J. L. (2016). One-step fabrication of near superhydrophobic aluminum surface by nanosecond laser ablation, *Appl. Surf. Sci.*, **374**, pp. 2–11.

187. Liu, Y., Liu, J., Li, S., Liu, J., Han, Z., and Ren, L. (2013). Biomimetic superhydrophobic Surface of high adhesion fabricated with micronano binary structure on aluminum alloy, *ACS Appl. Mater. Interfaces*, **5**, pp. 8907–8914.

188. Jagdheesh, R. (2014). Fabrication of a superhydrophobic Al_2O_3 surface using picosecond laser pulses, *Langmuir*, **30**, pp. 12067–12073.

189. Wang, X. Q., Ding, J. N., Yuan, N. Y., Wang, S. Y., Qiu, J. H., Kan, B., Guo, X. B., and Zhu, Y. Y. (2014). Switchable wettability of silicon micro-nano structures surface produced by femtosecond laser, *Key Eng. Mater.*, **609–610**, pp. 341–345.

190. Li, G. (2014). Superhydrophobicity of silicon-based microstructured surfaces, *Adv. Mater. Res.*, **989–994**, pp. 267–269.

191. Tang, M., Hong, M. H., Choo, Y. S., Tang, Z., and Chua, D. H. C. (2010). Super-hydrophobic transparent surface by femtosecond laser micro-patterned catalyst thin film for carbon nanotube cluster growth, *Appl. Phys. A*, **101**, pp. 503–508.

192. Moradi, S., Englezos, P., and Hatzikiriakos, S. G. (2014). Contact angle hysteresis of non-flattened-top micro/nanostructures, *Langmuir*, **30**, pp. 3274–3284.

193. Moradi, S., Kamal, S., Englezos, P., and Hatzikiriakos, S. G. (2013). Femtosecond laser irradiation of metallic surfaces: Effects of laser parameters on superhydrophobicity, *Nanotechnology*, **24**, pp. 415302/1–415302/12.

194. Toosi, S. F., Moradi, S., Ebrahimi, M., and Hatzikiriakos, S. G. (2016). Microfabrication of polymeric surfaces with extreme wettability using hot embossing, *Appl. Surf. Sci.*, **378**, pp. 426–434.

195. Wu, B., Zhou, M., Li, J., Ye, X., Li, G., and Cai, L. (2009). Superhydrophobic surfaces fabricated by microstructuring of stainless steel using a femtosecond laser, *Appl. Surf. Sci.*, **256**, pp. 61–66.
196. Kietzig, A.-M., Hatzikiriakos, S. G., and Englezos, P. (2009). Patterned superhydrophobic metallic surfaces, *Langmuir*, **25**, pp. 4821–4827.
197. Jagdheesh, R., Pathiraj, B., Karatay, E., Roemer, G. R. B. E., and Huis in't Veld, A. J. (2011). Laser-induced nanoscale superhydrophobic structures on metal surfaces, *Langmuir*, **27**, pp. 8464–8469.
198. Long, J., Fan, P., Zhong, M., Zhang, H., Xie, Y., and Lin, C. (2014). Superhydrophobic and colorful copper surfaces fabricated by picosecond laser induced periodic nanostructures, *Appl. Surf. Sci.*, **311**, pp. 461–467.
199. Liu, Y., Li, S., Niu, S., Cao, X., and Han, Z. (2016). Bio-inspired micro-nano structured surface with structural color and anisotropic wettability on Cu substrate, *Appl. Surf. Sci.*, **379**, pp. 230–237.
200. Peng, E., Tsubaki, A., Zuhlke, C. A., Wang, M., Bell, R., Lucis, M. J., Anderson, T. P., Alexander, D. R., Gogos, G., and Shield, J. E. (2016). Experimental explanation of the formation mechanism of surface mound-structures by femtosecond laser on polycrystalline $Ni_{60}Nb_{40}$, *Appl. Phys. Lett.*, **108**, pp. 031602/1–031602/5.
201. Kam, D. H., Bhattacharya, S., and Mazumder, J. (2012). Control of the wetting properties of an AISI 316L stainless steel surface by femtosecond laser-induced surface modification, *J. Micromech. Microeng.*, **22**, pp. 105019/1–105019/6.
202. Kwon, M. H., Shin, H. S., and Chu, C. N. (2014). Fabrication of a super-hydrophobic surface on metal using laser ablation and electrodeposition, *Appl. Surf. Sci.*, **288**, pp. 222–228.
203. Chen, T., Liu, H., Yang, H., Yan, W., Zhu, W., and Liu H. (2016). Biomimetic fabrication of robust self-assembly superhydrophobic surfaces with corrosion resistance properties on stainless steel substrates, *RSC Adv.*, **6**, pp. 43937–43949.
204. Zorba, V., Stratakis, E., Barberoglou, M., Spanakis, E., Tzanetakis, P., and Fotakis, C. (2008). Tailoring the wetting response of silicon surfaces via fs laser structuring, *Appl. Phys. A*, **93**, p. 819.
205. Barberoglou, M., Zorb, V., Stratakis, E., Spanakis, E., Tzanetakis, P., Anastasiadis, S. H., and Fotakis, C. (2009). Bio-inspired water repellent surfaces produced by ultrafast laser structuring of silicon, *Appl. Surf. Sci.*, **255**, pp. 5425–5429.

206. Barberoglou, M., Zorba, V., Pagozidis, A., Fotakis, C., and Stratakis, E. (2010). Electrowetting properties of micro/nanostructured black silicon, *Langmuir*, **26**, pp. 13007–13014.
207. Frysali, M. A., Papoutsakis, L., Kenanakis, G., and Anastasiadis, S. H. (2015). Functional surfaces with photocatalytic behavior and reversible wettability: ZnO coating on silicon spikes, *J. Phys. Chem. C*, **119**, pp. 25401–25407.
208. Steele, A., Nayak, B. K., Davis, A., and Gupta, M. C. (2013). Linear abrasion of a titanium superhydrophobic surface prepared by ultrafast laser microtexturing, *J. Micromech. Microeng.*, **23**, pp. 115012/1–115012/8.
209. Wang, D., Wang, Z., Zhang, Z., Yue, Y., Li, D., Qiu, R., and Maple, C. (2014). Both antireflection and superhydrophobicity structures achieved by direct laser interference nanomanufacturing, *J. Appl. Phys.*, **115**, pp. 233101/1–233101/6.
210. Zhang, G., Zhang, X., Li, M., and Su, Z. (2014). A surface with superoleophilic-to-superoleophobic wettability gradient, *ACS Appl. Mater. Interfaces*, **6**, pp. 1729–1733.
211. Feng, L., Yan, Z., Qiang, X., Liu, Y., and Wang, Y. (2016). Facile formation of superhydrophobic aluminum alloy surface and corrosion-resistant behavior, *Appl. Phys. A*, **122**, p. 165.
212. Feng, L., Zhang, H., Wang, Z., and Liu, Y. (2014). Superhydrophobic aluminum alloy surface: Fabrication, structure, and corrosion resistance, *Colloids Surf. A*, **441**, pp. 319–325.
213. Guo, M., Kang, Z., Li, W., and Zhang, J. (2014). A facile approach to fabricate a stable superhydrophobic film with switchable water adhesion on titanium surface, *Surf. Coat. Technol.*, **239**, pp. 227–232.
214. Florica, C., Preda, N., Costas, A., Zgura, I., and Enculescu, I. (2016). ZnO nanowires grown directly on zinc foils by thermal oxidation in air: Wetting and water adhesion properties, *Mater. Lett.*, **170**, pp. 156–159.
215. Pan, Q., Jin, H., and Wang, H. (2007). Fabrication of superhydrophobic surfaces on interconnected $Cu(OH)_2$ nanowires via solution-immersion, *Nanotechnology*, **18**, pp. 355605/1–355605/4.
216. Liang, W., Zhu, L., Li, W., and Liu, H. (2015). Facile fabrication of a flower-like $CuO/Cu(OH)_2$ nanorod film with tunable wetting transition and excellent stability, *RSC Adv.*, **5**, pp. 38100–38110.
217. Hu, J., Yuan, W., Yan, Z., Zhou, B., Tang, Y., and Li, Z. (2015). Fabricating an enhanced stable superhydrophobic surface on copper plates by introducing a sintering process, *Appl. Surf. Sci.*, **355**, pp. 145–152.

218. Lee, J.-Y., Han, J., Lee, J., Ji, S., and Yeo, J.-S. (2015). Hierarchical nanoflowers on nanograss structure for a non-wettable surface and a SERS substrate, *Nanoscale Res. Lett.*, **10**, p. 505.
220. Feng, L., Zhao, L., Qiang, X., Liu, Y., Sun, Z., and Wang, B. (2015). Fabrication of superhydrophobic copper surface with excellent corrosion resistance, *Appl. Phys. A*, **119**, pp. 75–83.
221. He, Z., Zhang, Z., and He, J. (2016). CuO/Cu based superhydrophobic and self-cleaning surfaces, *Scr. Mater.*, **118**, pp. 60–64.
222. Liu, N., Chen, Y., Lu, F., Cao, Y., Xue, Z., Li, K., Feng, L., and Wei, Y. (2013). Straightforward oxidation of a copper substrate produces an underwater superoleophobic mesh for oil/water separation, *ChemPhysChem*, **14**, pp. 3489–3494.
223. Liu, J., Wang, L., Guo, F., Hou, L., Chen, Y., Liu, J., Wang, N., Zhao, Y., and Jiang, L. (2016). Opposite and complementary: A superhydrophobic–superhydrophilic integrated system for high-flux, high-efficiency and continuous oil/water separation, *J. Mater. Chem. A*, **4**, pp. 4365–4370.
224. Zang, D., Wu, C., Zhu, R., Zhang, W., Yu, X., and Zhang, Y. (2013). Porous copper surfaces with improved superhydrophobicity under oil and their application in oil separation and capture from water, *Chem. Commun.*, **49**, pp. 8410–8412.
225. Li, J., Li, D., Li, W., She, H., Feng, H., and Hu, D. (2016). Facile fabrication of three-dimensional superhydrophobic foam for effective separation of oil and water mixture, *Mater. Lett.*, **171**, pp. 228–231.
226. Yu, C., Cao, M., Dong, Z., Wang, J., Li, K., and Jiang, L. (2016). Spontaneous and direction transportation of gas bubbles on superhydrophobic cones, *Adv. Funct. Mater.*, **26**, pp. 3236–3243.
227. Li, K., Zhang, J., Chen, J., Meng, G., Ding, Y., and Dai, Z. (2016). Centrifugation-assisted fog-collecting abilities of metal-foam structures with different surface wettabilities, *ACS Appl. Mater. Interfaces*, **8**, pp. 10005–10013.
228. Cho, H., Jeong, J., Kim, W., Choi, D., Lee, S., and Hwang, W. (2016). Conformable superoleophobic surfaces with multi-scale structures on polymer substrates, *J. Mater. Chem. A*, **4**, pp. 8272–8282.
229. Guo, Y., Wang, Z., and Wu, H. (2015). Facile fabrication of superhydrophobic film with high adhesion and the adhesive mechanism, *Appl. Phys. A*, **121**, pp. 1299–1303.
230. Chaudhary, A., and Barshilia, H. C. (2011). Nanometric multiscale rough CuO/Cu(OH)$_2$ superhydrophobic surfaces prepared by a facile one-step solution-immersion process: Transition to superhy-

drophilicity with oxygen plasma treatment, *J. Phys. Chem. C*, **115**, pp. 18213–18220.

231. Zhu, X., Zhang, Z., Men, X., Yang, J., and Xu, X. (2010). Rapid formation of superhydrophobic surfaces with fast response wettability transition, *ACS Appl. Mater. Interfaces*, **2**, pp. 3636–3641.

232. Yin, S., Wu, D., Yang, J., Lei, S., Kuang, T., and Zhu, B. (2011). Fabrication and surface characterization of biomimic superhydrophobic copper surface by solution-immersion and self-assembly, *Appl. Surf. Sci.*, **257**, pp. 8481–8485.

233. Zhang, Y., Yu, X., Zhou, Q., Chen, F., and Li, K. (2010). Fabrication of superhydrophobic copper surface with ultra-low water roll angle, *Appl. Surf. Sci.*, **256**, pp. 1883–1887.

234. Li, J., Liu, X., Ye, Y., Zhou, H., and Chen, J. (2011). Fabrication of superhydrophobic CuO surfaces with tunable water adhesion, *J. Phys. Chem. C*, **115**, pp. 4726–4729.

235. Pan, Q., Jin, H., and Wang, H. (2007). Fabrication of superhydrophobic surfaces on interconnected Cu(OH)$_2$ nanowires via solution-immersion, *Nanotechnology*, **18**, pp. 355605/1–355605/4.

236. Chen, X., Kong, L., Dong, D., Yang, G., Yu, L., Chen, J., and Zhang, P. (2009). Synthesis and characterization of superhydrophobic functionalized Cu(OH)$_2$ nanotube arrays on copper foil, *Appl. Surf. Sci.*, **255**, pp. 4015–4019.

237. Feng, J., Pang, Y., Qin, Z., Ma, R., and Yao, S. (2012). Why condensate drops can spontaneously move away on some superhydrophobic surfaces but not on others, *ACS Appl. Mater. Interfaces*, **4**, pp. 6618–6625.

238. Cheng, Z., Du, M., Lai, H., Zhang, N., and Sun, K. (2013). From petal effect to Lotus effect: A facile solution immersion process for the fabrication of super-hydrophobic surfaces with controlled adhesion, *Nanoscale*, **5**, pp. 2776–2783.

239. Zhu, H., and Guo, Z. (2015). Order separation of oil/water mixtures by superhydrophobic/superoleophilic Cu(OH)$_2$-thioled films, *Chem. Lett.*, **44**, pp. 1431–1433.

240. Liu, W., Xu, Q., Han, J., Chen, X., and Min, Y. (2016). A novel combination approach for the preparation of superhydrophobic surface on copper and the subsequent corrosion resistance, *Corros. Sci.*, **110**, pp. 105–113.

241. Guo, Z.-G., Fang, J., Hao, J.-C., Liang, Y.-M., and Liu, W.-M. (2006). A novel approach to stable superhydrophobic surfaces, *Chem Phys Chem*, **7**, pp. 1674–1677.

242. Cho, H., Lee, J., Lee, S., and Hwang, W. (2015). Durable superhydrophilic/phobic surfaces based on green patina with corrosion resistance, *Phys. Chem. Chem. Phys.*, **17**, pp. 6786–6793.
243. Kong, L., Chen, X., Yang, G., Yu, L., and Zhang, P. (2008). Preparation and characterization of slice-like $Cu_2(OH)_3NO_3$ superhydrophobic structure on copper foil, *Appl. Surf. Sci.*, **254**, pp. 7255–7258.
244. She, Z., Li, Q., Wang, Z., Li, L., Chen, F., and Zhou, J. (2012). Novel method for controllable fabrication of a superhydrophobic CuO surface on AZ91D magnesium alloy, *ACS Appl. Mater. Interfaces*, **4**, pp. 4348–4356.
245. Li, M., Su, Y., Hu, J., Yao, L., Wei, H., Yang, Z., and Zhang, Y. (2016). Hierarchically porous micro/nanostructured copper surfaces with enhanced antireflection and hydrophobicity, *Appl. Surf. Sci.*, **361**, pp. 11–17.
246. Lee, J.-Y., Pechook, S., Pokroy, B., Jeon, D.-J., Pokroy, B., and Yeo, J. S. (2014). Three-dimensional triple hierarchy formed by self-assembly of wax crystals on CuO nanowires for nonwettable surfaces, *ASC Appl. Mater. Interfaces*, **6**, pp. 4927–4934.
247. Lee, J. Y., Pechook, S., Pokroy, B., and Yeo, J. S. (2014). Multilevel hierarchy of fluorinated wax on CuO nanowires for superoleophobic surfaces, *Langmuir*, **30**, pp. 15568–15573.
248. Zhou, C., Cheng, J., Hou, K., Zhao, A., Pi, P., Wen, X., and Xu, S. (2016). Superhydrophilic and underwater superoleophobic titania nanowires surface for oil repellency and oil/water separation, *Chem. Eng. J.*, **301**, pp. 249–256.
249. Kim, A., Lee, C., Kim, H., and Kim, J. (2015). Simple approach to superhydrophobic nanostructured Al for practical antifrosting application based on enhanced self-propelled jumping droplets, *ACS Appl. Mater. Interfaces*, **7**, pp. 7206–7213.
250. Lakshmi, R. V., Bharathidasan, T., and Basu, B. J. (2013). Superhydrophobicity of AA2024 by a simple solution immersion technique, *Surf. Innovations*, **1**, pp. 241–247.
251. Song, H.-J., and Shen, X.-Q. (2010). Fabrication of functionalized aluminum compound petallike structure with superhydrophobic surface, *Surf. Interface Anal.*, **42**, pp. 165–168.
252. Peng, S., and Deng, W. (2015). A simple method to prepare superamphiphobic aluminum surface with excellent stability, *Colloids Surf. A*, **481**, pp. 143–150.
253. Pan, Q., and Cheng, Y. (2009). Superhydrophobic surfaces based on dandelion-like ZnO microspheres, *Appl. Surf. Sci.*, **255**, pp. 3904–3907.

254. Hou, X., Wang, L., Zhou, F., and Li, L. (2011). Fabrication of ZnO submicrorod films with water repellency by surface etching and hydrophobic modification, *Thin Solid Films*, **519**, pp. 7813–7816.
255. Wu, R., Chao, G., Jiang, H., Hu, Y., and Pan, A. (2015). The superhydrophobic aluminum surface prepared by different methods, *Mater. Lett.*, **142**, pp. 176–179.
256. Tsujii, K., Yamamoto, T., Onda, T., and Shibuichi, S. (1997). Super oil-repellent surfaces, *Angew. Chem. Int. Ed.*, **36**, pp. 1011–1012.
257. Shibuichi, S., Yamamoto, T., Onda, T., and Tsujii, K. (1998). Super water- and oil-repellent surfaces resulting from fractal structure, *J. Colloid Interface Sci.*, **208**, pp. 287–294.
258. Zheng, S., Li, C., Fu, Q., Hu, W., Xiang, T., Wang, Q., and Du, M. (2016). Development of stable superhydrophobic coatings on aluminum surface for corrosion-resistant, self-cleaning, anti-icing applications, *Mater. Des.*, **93**, pp. 261–270.
259. Jeong, C., and Choi, C.-H. (2012). Single-step direct fabrication of pillar-on-pore hybrid nanostructures in anodizing aluminum for superior superhydrophobic efficiency, *ACS Appl. Mater. Interfaces*, **4**, pp. 842–848.
260. Lu, Z., Wang, P., and Zhang, D. (2015). Super-hydrophobic film fabricated on aluminium surface as a barrier to atmospheric corrosion in a marine environment, *Corros. Sci.*, **91**, pp. 287–296.
261. Lee, W., Park, B. G., Kim, D. H., Ahn, D. J., Park, Y., Lee, S. H., and Lee, K. B. (2010). Nanostructure-dependent water-droplet adhesiveness change in superhydrophobic anodic aluminum oxide surfaces: From highly adhesive to self-cleanable, *Langmuir*, **26**, pp. 1412–1415.
262. Tang, K., Yu, J., Zhao, Y., Liu, Y., Wang, X., and Xu, R. (2006). Fabrication of super-hydrophobic and superoleophilic boehmite membranes from anodic alumina oxide film via a two-phase thermal approach, *J. Mater. Chem.*, **16**, pp. 1741–1745.
264. Kemell, M., Faerm, E., Leskelae, M., and Ritala, M. (2006). Transparent superhydrophobic surfaces by self-assembly of hydrophobic monolayers on nanostructured surfaces, *Phys. Stat. Sol.*, **203**, pp. 1453–1458.
265. Kim, D.-H., Kim, Y., Kim, B. M., Ko, J. S., Cho, C.-R., and Kim, J.-M. (2011). Uniform superhydrophobic surfaces using micro/nano complex structures formed spontaneously by a simple and cost-effective nonlithographic process based on anodic aluminum oxide technology, *J. Micromech. Microeng.*, **21**, pp. 045003/1–045003/8.

266. Lee, S., and Hwang, W. (2009). Ultralow contact angle hysteresis and no-aging effects in superhydrophobic tangled nanofiber structures generated by controlling the pore size of a 99.5% aluminum foil, *J. Micromech. Microeng.*, **19**, pp. 035019/1–035019/5.

267. Peng, S., Tian, D., Miao, X., Yang, X., and Deng, W. (2013). Designing robust alumina nanowires-on-nanopores structures: Superhydrophobic surfaces with slippery or sticky water adhesion, *J. Colloid Interface Sci.*, **409**, pp. 18–24.

268. Norek, M., and Krasiński, A. (2015). Controlling of water wettability by structural and chemical modification of porous anodic alumina (PAA): Towards super-hydrophobic surfaces, *Surf. Coat. Technol.*, **276**, pp. 464–470.

269. Zhang, H., Yin, L., Shi, S., Liu, X., Wang, Y., and Wang, F. (2015). Facile and fast fabrication method for mechanically robust superhydrophobic surface on aluminum foil, *Microelectron. Eng.*, **141**, pp. 238–242.

270. Zhang, H., Yin, L., Li, L., Shi, S., Wang, Y., and Liu, X. (2016). Wettability and drag reduction of a superhydrophobic aluminum surface, *RSC Adv.*, **6**, pp. 14034–14041.

271. Lee, S., Kim, W., Lee, S., Shim, S., and Choi, D. (2015). Controlled transparency and wettability of large-area nanoporous anodized alumina on glass, *Scr. Mater.*, **104**, pp. 29–32.

272. Kim, Y., Lee, S., Cho, H., Park, B., Kim, D., and Hwang, W. (2012). Robust superhydrophilic/hydrophobic surface based on self-aggregated Al_2O_3 nanowires by single-step anodization and self-assembly Method, *ACS Appl. Mater. Interfaces*, **4**, pp. 5074–5078.

273. Fujii, T., Aoki, Y., and Habazaki, H. (2011). Fabrication of super-oil-repellent dual pillar surfaces with optimized pillar intervals, *Langmuir*, **27**, pp. 11752–11756.

274. Wang, G., Liu, S., Wei, S., Liu, Y., Lian, J., and Jiang, Q. (2016). Robust superhydrophobic surface on Al substrate with durability, corrosion resistance and ice-phobicity, *Sci. Rep.*, **6**, p. 20933.

275. Wu, Y., Zhao, W., Wang, W., and Sui, W. (2016). Fabricating binary anti-corrosion structures containing superhydrophobic surfaces and sturdy barrier layers for Al alloys, *RSC Adv.*, **6**, pp. 5100–5110.

276. Nakayama, K., Tsuji, E., Aoki, Y., Park, S.-G., and Habazaki, H. (2016). Control of surface wettability of aluminum mesh with hierarchical surface morphology by monolayer coating: From superoleophobic to superhydrophilic, *J. Phys. Chem. C*, **120**, pp. 15684–15690.

277. Ganne, A., Lebed, V. O., and Gavrilov, A. I. (2016). Combined wet chemical etching and anodic oxidation for obtaining the superhydrophobic meshes with anti-icing performance, *Colloids Surf. A*, **499**, pp. 150–1553.

278. Liu, Y., Cao, H., Chen, S., and Wang, D. (2015). Ag nanoparticle-loaded hierarchical superamphiphobic surface on an Al substrate with enhanced anticorrosion and antibacterial properties, *J. Phys. Chem. C*, **119**, pp. 25449–25456.

279. Song, T., Liu, Q., Zhang, M., Chen, R., Takahashi, K., Jing, X., Liu, L., and Wang, J. (2015). Multiple sheet-layered super slippery surfaces based on anodic aluminium oxide and its anticorrosion property, *RSC Adv.*, **5**, pp. 70080–70085.

280. Wang, X. J., Song, W., Li, Z. S., and Cong, Q. (2012). Fabrication of superhydrophobic AAO-Ag multilayer mimicking dragonfly wings, *Chin. Sci. Bull.*, **57**, pp. 4635–4640.

281. Tian, D., Zhai, J., Song, Y., and Jiang, L. (2011). Photoelectric cooperative induced wetting on aligned-nanopore arrays for liquid reprography, *Adv. Funct. Mater.*, **21**, pp. 4519–4526.

282. Ahmad, N. A., Leo, C. P., and Ahmad, A. L. (2013). Synthesis of superhydrophobic alumina membrane: Effects of sol–gel coating, steam impingement and water treatment, *Appl. Surf. Sci.*, **284**, pp. 556–564.

283. Rana, K., Kucukayan-Dogu, G., and Bengu, E. (2012). Growth of vertically aligned carbon nanotubes over self-ordered nano-porous alumina films and their surface properties, *Appl. Surf. Sci.*, **258**, pp. 7112–7117.

284. Li, Y., Li, S., Zhang, Y., Yu, M., and Liu, J. (2015). Fabrication of superhydrophobic layered double hydroxides films with different metal cations on anodized aluminum 2198 alloy, *Mater. Lett.*, **142**, pp. 137–140.

285. Zhang, F., Zhao, L., Chen, H., Xu, S., Evans, D. G., and Duan, X. (2008). Corrosion resistance of superhydrophobic layered double hydroxide films on aluminum, *Angew. Chem. Int. Ed.*, **47**, pp. 2466–2469.

286. Chen, T., Xu, S., Zhang, F., Evans, D. G., and Duan, X. (2009). Formation of photo- and thermo-stable layered double hydroxide films with photo-responsive wettability by intercalation of functionalized azobenzenes, *Chem. Eng. Sci.*, **64**, pp. 4350–4357.

287. Chen, H., Zhang, F., Fu, S., and Duan, X. (2006). In situ microstructure control of oriented layered double hydroxide monolayer films

with curved hexagonal crystals as superhydrophobic materials, *Adv. Mater.*, **18**, pp. 3089–3093.

288. Kong, J.-H., Kim, T.-H., Kim, J. H., Park, J.-K., Lee, D.-W., Kim, S.-H., and Kim, J.-M. (2014). Highly flexible, transparent and self-cleanable superhydrophobic films prepared by a facile and scalable nanopyramid formation technique, *Nanoscale*, **6**, pp. 1453–1461.

289. Kim, T.-H., Ha, S.-H., Jang, N.-S., Kim, J., Kim, J. H., Park, J.-K., Lee, D.-W., Lee, J., Kim, S.-H., and Kim, J.-M. (2015). Simple and cost-effective fabrication of highly flexible, transparent superhydrophobic films with hierarchical surface design, *ACS Appl. Mater. Interfaces*, **7**, pp. 5289–5295.

290. Neto, C., Joseph, K. R., and Brant, W. R. (2009). On the superhydrophobic properties of nickel nanocarpets, *Phys. Chem. Chem. Phys.*, **11**, pp. 9537–9544.

291. Liu, M., Liu, X., Wang, J., Wei, Z., and Jiang, L. (2010). Electromagnetic synergetic actuators based on polypyrrole/Fe_3O_4 hybrid nanotube arrays, *Nano Res.*, **3**, pp. 670–675.

292. Chen, X., Chen, G., Ma, Y., Li, X., Jiang, L., and Wang, F. (2006). Conductive super-hydrophobic surfaces of polyaniline modified porous anodic alumina membranes, *J. Nanosci. Nanotechnol.*, **6**, pp. 783–786.

293. Bok, H.-M., Shin, T.-Y., and Park, S. (2008). Designer binary nanostructures toward water slipping superhydrophobic surfaces, *Chem. Mater.*, **20**, pp. 2247–2251.

294. Cui, G., Xu, W., Zhou, X., Xiao, X., Jiang, L., and Zhu, D. (2006). Rose-like superhydrophobic surface based on conducting dmit salt. *Colloids Surf. A*, **272**, pp. 63–67.

295. Daglar, B., Khudiyev, T., Birlik Demirel, G., Buyukserine, F., and Bayindir, M. (2013). Soft biomimetic tapered nanostructures for large-area antireflective surfaces and SERS sensing, *J. Mater. Chem. C*, **1**, pp. 7842–7848.

296. Zhang, L., Zhou, Z., Cheng, B., DeSimone, J. M., and Samulski, E. T. (2006). Superhydrophobic behavior of a perfluoropolyether Lotus-leaf-like topography, *Langmuir*, **22**, pp. 8576–8580.

297. Sheng, X., and Zhang, J. (2009). Superhydrophobic behaviors of polymeric surfaces with aligned nanofibers, *Langmuir*, **25**, pp. 6916–6922.

298. Puukilainen, E., Rasilainen, T., Suvanto, M., and Pakkanen, T. A. (2007). Superhydrophobic polyolefin surfaces: Controlled micro- and nanostructures, *Langmuir*, **23**, pp. 7263–7268.

299. Feng, L., Li, S., Li, H., Zhai, J., Song, Y., Jiang, L., and Zhu, D. (2002). Super-hydrophobic surface of aligned polyacrylonitrile nanofibers, *Angew. Chem. Int. Ed.*, **41**, pp. 1221–1224.
300. Feng, L., Song, Y., Zhai, J., Liu, B., Xu, J., Jiang, L., and Zhu, D. (2003). Creation of a superhydrophobic surface from an amphiphilic polymer, *Angew. Chem. Int. Ed.*, **42**, pp. 800–802.
301. Eliseev, A. A., Petukhov, D. I., Buldakov, D. A., Ivanov, R. P., Napolskii, K. S., Lukashin, A. V., and Tret'yakov Y. D. (2010). Morphology modification of the surface of polymers by the replication of the structure of anodic aluminum oxide, *JETP Lett.*, **92**, pp. 453–456.
302. Cheng, Z., Gao, J., and Jiang, L. (2010). Tip geometry controls adhesive states of superhydrophobic surfaces, *Langmuir*, **26**, pp. 8233–8238.
303. Jin, M., Feng, X., Feng, L., Sun, T., Zhai, J., Li, T., and Jiang, L. (2005). Superhydrophobic aligned polystyrene nanotube films with high adhesive force, *Adv. Mater.*, **17**, pp. 1977–1981.
304. Hong, X., Gao, X., and Jiang, L. (2007). Application of superhydrophobic surface with high adhesive force in no lost transport of superparamagnetic microdroplet, *J. Am. Chem. Soc.*, **129**, pp. 1478–1479.
305. Lovera, P., Creedon, N., Alatawi, H., Mitchell, M., Burke, M., Quinn, A. J., and O'Riordan, A. (2014). Low-cost silver capped polystyrene nanotube arrays as super-hydrophobic substrates for SERS applications, *Nanotechnology*, **25**, pp. 175502/1–175502/6.
306. Chu, D., Nemoto, A., and Ito, H. (2015). Biomimetic superhydrophobic polymer surfaces by replication of hierarchical structures fabricated using precision tooling machine and anodized aluminum oxidation, *Microsyst. Technol.*, **21**, pp. 123–130.
307. Tian, W., Xu, Y., Huang, L., Yung, K.-L., Xie, Y., and Chen, W. (2011). β-cyclodextrin and its hyperbranched polymers-induced micro/nanopatterns and tunable wettability on polymer surfaces, *Nanoscale*, **3**, pp. 5147–5155.
308. Hong, D., You, I., Lee, H., Lee, S.-G., Choi, I. S., and Kang, S. M. (2013). Polydopamine circle-patterns on a superhydrophobic AAO surface: Water-capturing property, *Bull. Korean Chem. Soc.*, **34**, pp. 3141–3142.
309. Lee, W., Park, B. G., and Lee, K. B. (2010). Fabrication of superhydrophobic surface of a alumina with nanoporous structure, *Bull. Korean Chem. Soc.*, **31**, pp. 1833–1834.
310. Cho, W. K., and Choi, I. S. (2008). Fabrication of hairy polymeric polymer films inspired by geckos: Wetting and high adhesion properties, *Adv. Funct. Mater.*, **18**, pp. 1089–1096.

311. Peng, S., and Deng, W. (2014). A facile approach for preparing biomimetic polymer macroporous structures with petal or lotus effects, *New J. Chem.*, **38**, pp. 1011–1018.
312. Mozalev, A., Habazaki, H., and Hubálek, J. (2012). The superhydrophobic properties of self-organized microstructured surfaces derived from anodically oxidized Al/Nb and Al/Ta metal layers, *Electrochim. Acta*, **82**, pp. 90–97.
313. Wang, J., and Lin, Z. (2009). Anodic formation of ordered TiO_2 nanotube arrays: Effects of electrolyte temperature and anodization potential, *J. Phys. Chem. C*, **113**, pp. 4026–4030.
314. Lai, Y., Pan, F., Xu, C., Fuchs, H., and Chi, L. (2012). In situ surface-modification-induced superhydrophobic patterns with reversible wettability and adhesion, *Adv. Mater.*, **25**, pp. 1682–1686.
315. Lai, Y., Lin, L., Pan, F., Huang, J., Song, R., Huang, Y., Lin, C., Fuchs, H., and Chi, L. (2013). Bioinspired patterning with extreme wettability contrast on TiO_2 nanotube array surface: A versatile platform for biomedical applications, *Small*, **9**, pp. 2945–2953.
316. Li, H., Lai, Y., Huang, J., Tang, Y., Yang, L., Chen, Z., Zhang, K., Wang, X., and Tan, L. P. (2015). Multifunctional wettability patterns prepared by laser processing on superhydrophobic TiO_2 nanostructured surfaces, *J. Mater. Chem. B*, **3**, pp. 342–347.
317. Lai, Y., Huang, Y., Wang, H., Huang, J., Chen, Z., and Lin, C. (2010). Selective formation of ordered arrays of octacalcium phosphate ribbons on TiO_2 nanotube surface by template-assisted electrodeposition, *Colloids Surf. B*, **76**, pp. 117–122.
318. Dong, J., Ouyang, X., Han, J., Qiu, W., and Gao, W. (2014). Superhydrophobic surface of TiO_2 hierarchical nanostructures fabricated by Ti anodization, *J. Colloid Interface Sci.*, **420**, pp. 97–100.
319. Fan, X., Li, X., Tian, D., Zhai, J., and Jiang, L. (2012). Optoelectrowettability conversion on superhydrophobic CdS QDs sensitized TiO_2 nanotubes, *J. Colloid Interface Sci.*, **366**, pp. 1–7.
320. Barthwal, S., Kim, Y. S., and Lim, S.-H. (2013). Fabrication of amphiphobic surface by using titanium anodization for large-area three-dimensional substrates, *J. Colloid Interface Sci.*, **400**, pp. 123–129.
321. Wang, D., Wang, X., Liu, X., and Zhou, F. (2010). Engineering a titanium surface with controllable oleophobicity and switchable oil adhesion, *J. Phys. Chem. C*, **114**, pp. 9938–9944.

322. Lai, Y.-K., Tang, Y.-X., Huang, J.-Y., Pan, F., Chen, Z., Zhang, K.-Q., Fuchs, H., and Chi, L.-F. (2013). Bioinspired TiO$_2$ nanostructure films with special wettability and adhesion for droplets manipulation and patterning, *Sci. Rep.*, **3**, p. 3009.
323. Lai, Y., Lin, C., Huang, J., Zhuang, H., Sun, L., and Nguyen, T. (2008). Markedly controllable adhesion of superhydrophobic spongelike nanostructure TiO$_2$ films, *Langmuir*, **24**, pp. 3867–3873.
324. Huang, J.-Y., Lai, Y.-K., Pan, F., Yang, L., Wang, H., Zhang, K.-Q., Fuchs, H., and Chi, L.-F. (2014). Multifunctional superamphiphobic TiO$_2$ nanostructure surfaces with facile wettability and adhesion engineering, *Small*, **10**, pp. 4865–4873.
325. Li, S.-Y., Li, Y., Wang, J., Nan, Y.-G., Ma, B.-H., Liu, Z.-L., and Gu, J.-X. (2016). Fabrication of pinecone-like structure superhydrophobic surface on titanium substrate and its self-cleaning property, *Chem. Eng. J.*, **290**, pp. 82–90.
326. Wu, X., and Shi, G. (2006). Production and characterization of stable superhydrophobic surfaces based on copper hydroxide nanoneedles mimicking the legs of water striders, *J. Phys. Chem. B*, **110**, pp. 11247–11252.
327. Xiao, F., Yuan, S., Liang, B., Li, G., Pehkonen, S. O., and Zhang, T. (2015). Superhydrophobic CuO nanoneedle-covered copper surfaces for anticorrosion, *J. Mater. Chem. A*, **3**, pp. 4374–4388.
328. Jiang, W., He, J., Xiao, F., Yuan, S., Lu, H., and Liang, B. (2015). Preparation and antiscaling application of superhydrophobic anodized CuO nanowire surfaces, *Ind. Eng. Chem. Res.*, **54**, pp. 6874–6883.
329. She, Z., Li, Q., Wang, Z., Li, L., Chen, F., and Zhou, J. (2012). Novel method for Controllable fabrication of a superhydrophobic CuO surface on AZ91D magnesium alloy, *ACS Appl. Mater. Interfaces*, **4**, pp. 4348–4356.
330. La, D.-D., Nguyen, T. A., Lee, S., Kim, J. W., and Kim, Y. S. (2011). A stable superhydrophobic and superoleophilic Cu mesh based on copper hydroxide nanoneedle arrays, *Appl. Surf. Sci.*, **257**, pp. 5705–5710.
331. Jiang, W., He, Jian, Mao, M., Yuang, S., Lu, H., and Liang, B. (2016). Preparation of superhydrohobic Cu mesh and its application in rolling-spheronization granulation, *Ind. Eng. Chem. Res.*, **55**, pp. 5545–5555.
332. He, S., Zheng, M., Yao, L., Yuan, X., Li, M., Ma, L., and Shen, W. (2010). Preparation and properties of ZnO nanostructures by electrochemical anodization method, *Appl. Surf. Sci.*, **256**, pp. 2557–2562.

333. Ramirez-Canon, A., Miles, D. O., Cameron, P. J., and Mattia, D. (2013). Zinc oxide nanostructured films produced via anodization: A rational design approach, *RSC Adv.*, **3**, pp. 25323–25330.
334. Yang, Y., Liu, J., Li, C., Fu, L., Huang, W., and Li, Z. (2012). Fabrication of pompon-like and flower-like SnO microspheres comprised of layered nanoflakes by anodic electrocrystallization, *Electrochim. Acta*, **72**, pp. 94–100.
335. Yang, S., Habazaki, H., Fujii, T., Aoki, Y., Skeldon, P., and Thompson, G. E. (2011). Control of morphology and surface wettability of anodic niobium oxide microcones formed in hot phosphate–glycerol electrolytes, *Electrochim. Acta*, **56**, pp. 7446–7453.
336. Gu, C., and Zhang, T.-Y. (2008). Electrochemical synthesis of silver polyhedrons and dendritic films with superhydrophobic surfaces, *Langmuir*, **24**, pp. 12010–12016.
337. Che, P., Liu, W., Chang, X., Wang, A., and Han, Y. (2016). Multifunctional silver film with superhydrophobic and antibacterial properties, *Nano Res.*, 9, pp. 442–450.
338. Liu, G., Duan, G., Jia, L., Wang, J., Wang, H., Cai, W., and Li, Y. (2013). Fabrication of self-standing silver nanoplate arrays by seed-decorated electrochemical route and their structure-induced properties, *J. Nanomater.*, **2013**, pp. 365947.
339. Yang, S., Hricko, P. J., Huang, P.-H., Li, S., Zhao, Y., Xie, Y., Guo, F., Wang, L., and Huang, T. J. (2014). Superhydrophobic surface enhanced Raman scattering sensing using Janus particle arrays realized by site-specific electrochemical growth, *J. Mater. Chem. C*, **2**, pp. 542–547.
340. Wu, Y., Liu, K., Su, B., and Jiang, L. (2014). Superhydrophobicity-mediated electrochemical reaction along the solid–liquid–gas triphase interface: Edge-growth of gold architectures, *Adv. Mater.*, **26**, pp. 1124–1128.
341. Wang, L., Guo, S., Hu, X., and Dong, S. (2008). Facile electrochemical approach to fabricate hierarchical flowerlike gold microstructures: Electrodeposited superhydrophobic surface, *Electrochem. Commun.*, **10**, pp. 95–99.
342. Li, Y., and Shi, G. (2005). Electrochemical growth of two-dimensional gold nanostructures on a thin polypyrrole film modified ITO electrode, *J. Phys. Chem. B*, **109**, pp. 23787–23793.
343. Wang, J., Duan, G., Li, Y., Liu, G., Dai, Z., Zhang, H., and Cai, W. (2013). An invisible template method toward gold regular arrays of nanoflowers by electrodeposition, *Langmuir*, **29**, pp. 3512–3517.

344. Choi, S., Kweon, S., and Kim, J. (2015). Electrodeposition of Pt nanostructures with reproducible SERS activity and superhydrophobicity, *Phys. Chem. Chem. Phys.*, **17**, pp. 23547–23553.

345. Jeong, H., and Kim, J. (2015). Electrodeposition of nanoflake Pd structures: Structure-dependent wettability and SERS activity, *ACS Appl. Mater. Interfaces*, **7**, pp. 7129–7135.

346. Liu, Q., Tang, Y., Luo, W., Fu, T., and Yuan, W. (2015). Fabrication of superhydrophilic surface on copper substrate by electrochemical deposition and sintering process, *Chin. J. Chem. Eng.*, **23**, pp. 1200–1205.

347. Rahmawan, Y., Xu, L., and Yang, S. (2016). Self-assembled superhydrophobic multilayer films with corrosion resistance on copper substrate, *RSC Adv.*, **6**, pp. 2379–2386.

348. Liu, Y., Zhang, K., Yao, W., Zhang, C., Han, Z., and Ren, L. (2016). A facile electrodeposition process for the fabrication of superhydrophobic and superoleophilic copper mesh for efficient oil–water separation, *Ind. Eng. Chem. Res.*, **55**, pp. 2704–2712.

349. Xu, N., Sarkar, D. K., Chen, X. G., Zhang, H., and Tong, W. (2016). Superhydrophobic copper stearate/copper oxide thin films by a simple one-step electrochemical process and their corrosion resistance properties, *RSC Adv.*, **6**, pp. 35466–35478.

350. Wang, F., Lei, S., Xue, M., Ou, J., and Li, W. (2014). In situ separation and collection of oil from water surface via a novel superoleophilic and superhydrophobic oil containment boom, *Langmuir*, **30**, pp. 1281–1289.

351. Liu, L., Liu, W., Chen, R., Li, X., and Xie, X. (2015). Hierarchical growth of Cu zigzag microstrips on Cu foil for superhydrophobicity and corrosion resistance, *Chem. Eng. J.*, **281**, pp. 804–812.

352. Haghdoost, A., and Pitchumani, R. (2014). Fabricating superhydrophobic surfaces via a two-step electrodeposition technique, *Langmuir*, **30**, pp. 4183–4191.

353. Deng, Y., Ling, H., Feng, X., Hang, T., and Li, M. (2015). Electrodeposition and characterization of copper nanocone structures, *Cryst Eng Comm*, **17**, pp. 868–876.

354. Wang, N., Yuan, Y., Wu, Y., Hang, T., and Li, M. (2015). Wetting transition of the caterpillar-like superhydrophobic Cu/Ni-Co hierarchical structure by heat treatment, *Langmuir*, **31**, pp. 10807–10812.

355. Wang, H., Wang, N., Hang, T., and Li, M. (2016). Morphologies and wetting properties of copper film with 3D porous micro-nano

hierarchical structure prepared by electrochemical deposition, *Appl. Surf. Sci.*, **372**, pp. 7–12.

356. Guo, M., Liu, M., Zhao, W., Xia, Y., Huang, W., and Li, Z. (2015). Rapid fabrication of SERS substrate and superhydrophobic surface with different micro/nano-structures by electrochemical shaping of smooth Cu surface, *Appl. Surf. Sci.*, **353**, pp. 1277–1284.

357. Li, Y., Jia, W.-Z., Song, Y.-Y., and Xia, X.-H. (2007). Superhydrophobicity of 3D porous copper films prepared using the hydrogen bubble dynamic template, *Chem. Mater.*, **19**, pp. 5758–5764.

358. Yao, X., Xu, L., and Jiang, L. (2010). Fabrication and characterization of superhydrophobic surfaces with dynamic stability, *Adv. Funct. Mater.*, **20**, pp. 3343–3349.

359. Mahajan, M., Bhargava, S. K., and O'Mullane, A. P. (2013). Electrochemical formation of porous copper 7,7,8,8-tetracyanoquinodimethane and copper 2,3,5,6-tetrafluoro-7,7,8,8-tetracyanoquinodimethane honeycomb surfaces with superhydrophobic properties, *Electrochim. Acta*, **101**, pp. 186–195.

360. Meng, K., Jiang, Y., Jiang, Z., Lian, J., and Jiang, Q. (2014). Cu surfaces with controlled structures: From intrinsically hydrophilic to apparently superhydrophobic, *Appl. Surf. Sci.*, **290**, pp. 320–326.

361. Zhao, W., Guo, M., Xia, Y., Huang, W., and Li, Z. (2017). Preparation of porous Cu_2O/Cu_2S coatings for superhydrophobicity by fast electrochemical treatment of smooth copper substrates, *Surf. Coat. Technol.*, in press.

362. Lee, J. M., Bae, K. M., Jung, K. K., Jeong, J. H., and Ko, J. S. (2014). Creation of microstructured surfaces using Cu–Ni composite electrodeposition and their application to superhydrophobic surfaces, *Appl. Surf. Sci.*, **289**, pp. 14–20.

363. Zhang, J., Baró, M. D., Pellicer, E., and Sort, J. (2014). Electrodeposition of magnetic, superhydrophobic, non-stick, two-phase Cu–Ni foam films and their enhanced performance for hydrogen evolution reaction in alkaline water media, *Nanoscale*, **6**, pp. 12490–12499.

364. Jamali-Sheini, F., Yousefi, R., Bakr, N. A., Mahmoudian, M. R., Singh, J., and Huang, N. M. (2014). Electrodeposition of Cu–ZnO nanocomposites: Effect of growth conditions on morphologies and surface properties, *Mater. Sci. Semiconductor Process.*, **27**, pp. 507–514.

365. Jamali-Sheini, F., and Yousefi, R. (2013). Electrochemical synthesis and surface characterization of hexagonal Cu–ZnO nano-funnel tube films, *Ceramics Int.*, **39**, pp. 3715–3720.

366. Liu, Y., Yin, X., Zhang, J., Yu, S., Han, Z., and Ren, L. (2014). A electro-deposition process for fabrication of biomimetic superhydrophobic surface and its corrosion resistance on magnesium alloy, *Electrochim. Acta*, **125**, pp. 395–403.
367. Su, F., and Yao, K. (2014). Facile fabrication of superhydrophobic surface with excellent mechanical abrasion and corrosion resistance on copper substrate by a novel method, *ACS Appl. Mater. Interfaces*, **6**, pp. 8762–8770.
368. Zhang, E., Liu, Y., Yu, J., Lv, T., and Li, L. (2015). Fabrication of hierarchical gecko-inspired microarrays using a three-dimensional porous nickel oxide template, *J. Mater. Chem. B*, **3**, pp. 6571–6575.
369. Gu, C., and Tu, J. (2011). One-step fabrication of nanostructured Ni film with Lotus effect from deep eutectic solvent, *Langmuir*, **27**, pp. 10132–10140.
370. Chang, Y.-H., Huang, Y.-T., Lo, M. K., Lin, C.-F., Chen, C.-M., and Feng, S.-P. (2014). Electrochemical fabrication of transparent nickel hydroxide nanostructures with tunable superhydrophobicity/superhydrophilicity for 2D microchannels application, *J. Mater. Chem. A*, **2**, pp. 1985–1990.
371. Chang, Y.-H., Hau, N. Y., Liu, C., Huang, Y.-T., Li, C.-C., Shih, K., and Feng, S.-P. (2014). A short-range ordered–disordered transition of a NiOOH/Ni(OH)$_2$ pair induces switchable wettability, *Nanoscale*, **6**, pp. 15309–15315.
372. Esmailzadeh, S., Khorsand, S., Raeissi, K., and Ashrafizadeh, F. (2015). Microstructural evolution and corrosion resistance of superhydrophobic electrodeposited nickel films, *Surf. Coat. Technol.*, **283**, pp. 337–346.
373. Khorsand, S., Raeissi, K., and Ashrafizadeh, F. (2014). Corrosion resistance and long-term durability of super-hydrophobic nickel film prepared by electrodeposition process, *Appl. Surf. Sci.*, **305**, pp. 498–505.
374. Hashemzadeh, M., Raeissi, K., Ashrafizadeh, F., and Khorsand, S. (2015). Effect of ammonium chloride on microstructure, superhydrophobicity and corrosion resistance of nickel coatings, *Surf. Coat. Technol.*, **283**, pp. 318–328.
375. Hu, F., Xu, P., Wang, H., Kang, U. B., Hu, A., and Li, M. (2015). Superhydrophobic and anti-corrosion Cu microcones/Ni–W alloy coating fabricated by electrochemical approaches, *RSC Adv.*, **5**, pp. 103863–103868.

376. Lee, J. M., Jung, K. K., and Ko, J. S. (2016). Effect of NaCl in a nickel electrodeposition on the formation of nickel nanostructure, *J. Mater. Sci.*, **51**, pp. 3036–3044.
377. Liang, J., Li, D., Wang, D., Liu, K., and Chen, L. (2014). Preparation of stable superhydrophobic film on stainless steel substrate by a combined approach using electrodeposition and fluorinated modification, *Appl. Surf. Sci.*, **293**, pp. 265–270.
378. Mo, X., Wu, Y., Zhang, J., Hang, T., and Li, M. (2015). Bioinspired multifunctional Au nanostructures with switchable adhesion, *Langmuir*, **31**, pp. 10850–10858.
379. Khorsand, S., Raeissi, K., Ashrafizadeh, F., and Arenas, M. A. (2015). Super-hydrophobic nickel–cobalt alloy coating with micro-nano flower-like structure, *Chem. Eng. J.*, **273**, pp. 638–646.
380. Khorsand, S., Raeissi, K., Ashrafizadeh, F., and Arenas, M. A. (2015). Relationship between the structure and water repellency of nickel–cobalt alloy coatings prepared by electrodeposition process, *Surf. Coat. Technol.*, **276**, pp. 296–304.
381. Khorsand, S., Raeissi, K., Ashrafizadeh, F., Arenas, M. A., and Conde, A. (2016). Corrosion behaviour of super-hydrophobic electrodeposited nickel–cobalt alloy film, *Appl. Surf. Sci.*, **364**, pp. 349–357.
382. Sivasakthi, P., Ramesh Bapu, G. N. K., Chandrasekaran, M., and Sreejakumari, S. S. (2016). Synthesis and super-hydrophobic Ni-ITO nanocomposite with pine-cone and spherical shaped micro-nanoarchitectures by pulse electrodeposition and its electrocatalytic application, *RSC Adv.*, **6**, pp. 44766–44773.
383. Escobar, A. M., Llorca-Isern, N., and Rius-Ayra, O. (2016). Identification of the mechanism that confers superhydrophobicity on 316L stainless steel, *Mater. Charact.*, **111**, pp. 162–169.
384. Fan, Y., He, Y., Luo, P., Chen, X., and Liu, B. (2016). A facile electrodeposition process to fabricate corrosion-resistant super hydrophobic surface on carbon steel, *Appl. Surf. Sci.*, **368**, pp. 435–442.
385. Chen, Z., Hao, L., and Chen, C. (2012). Simultaneous fabrication of superhydrophobic coatings on cathodic and anodic copper surfaces with micro/nano-structures, *ECS Electrochem. Lett.*, **1**, pp. D21–D23.
386. Chen, Z., Li, F., Hao, L., Chen, A., and Kong, Y. (2011). One-step electrodeposition process to fabricate cathodic superhydrophobic surface, *Appl. Surf. Sci.*, **258**, pp. 1395–1398.

387. Qiu, R., Wang, P., Zhang, D., and Wu, J. (2011). One-step preparation of hierarchical cobalt structure with inborn superhydrophobic effect, *Colloids Surf. A*, **377**, pp. 144–149.
388. Qiu, R., Zhang, D., Wang, P., Zhang, X. L., and Kang, Y. S. (2011). Tunable electrochemical preparation of cobalt micro/nanostructures and their morphology-dependent wettability property, *Electrochim. Acta*, **58**, pp. 699–706.
389. Suz, F., Yao, K., Liu, C., and Huang, P. (2013). Rapid fabrication of corrosion resistant and superhydrophobic cobalt coating by a one-step electrodeposition, *J. Electrochem. Soc.*, **160**, pp. D59–D599.
390. Yanpeng, X., Taleb, A., and Jegou, P. (2013). Electrodeposition of cobalt films with an oriented fir tree-like morphology with adjustable wetting properties using a self-assembled gold nanoparticle modified HOPG electrode, *J. Mater. Chem. A*, **1**, pp. 11580–11588.
391. Xiao, H., Hu, A., Hang, T., and Li, M. (2015). Electrodeposited nanostructured cobalt film and its dual modulation of both superhydrophobic property and adhesiveness, *Appl. Surf. Sci.*, **324**, pp. 319–323.
392. Li, M., Zhai, J., Liu, H., Song, Y., Jiang, L., and Zhu, D. (2003). Electrochemical deposition of conductive superhydrophobic zinc oxide thin films, *J. Phys. Chem. B*, **107**, pp. 9954–9957.
393. Zhang, B., Lu, S., Xu, W., and Cheng, Y. (2016). Controllable wettability and morphology of electrodeposited surfaces on zinc substrates, *Appl. Surf. Sci.*, **360**, pp. 904–914.
394. He, G., Lu, S., Xu, W., Szunerits, S., Boukherroub, R., and Zhang, H. (2015). Controllable growth of durable superhydrophobic coatings on a copper substrate via electrodeposition, *Phys. Chem. Chem. Phys.*, **17**, pp. 10871–10880.
395. Chen, L.-Y., Lai, C.-H., Wu, P.-W., and Fan S.-K. (2011). Electrowetting of superhydrophobic ZnO inverse opals, *J. Electrochem. Soc.*, **158**, pp. P93–P99.
396. Pauporte, T., Bataille, G., Joulaud, L., and Vermersch, F. J. (2010). Well-aligned ZnO nanowire arrays prepared by seed-layer-free electrodeposition and their Cassie–Wenzel transition after hydrophobization, *J. Phys. Chem. C*, **114**, pp. 194–202.
397. Badre, C., Pauporte, T., Turmine, M., Dubot, P., and Lincot, D. (2008). Water-repellent ZnO nanowires films obtained by octadecylsilane self-assembled monolayers, *Phys. E*, **40**, pp. 2454–2456.
398. Badre, C., Dubot, P., Lincot, D., Pauporte, T., and Turmine, M. (2007). Effects of nanorod structure and conformation of fatty acid self-

assembled layers on superhydrophobicity of zinc oxide, *J. Colloid Interface Sci.*, **316**, pp. 233–237.

399. Badre, C., and Pauporté, T. (2009). Nanostructured ZnO-based surface with reversible electrochemically adjustable wettability, *Adv. Mater.*, **21**, pp. 697–701.

400. Hsieh, C.-T., Yang, S.-Y., and Lin, J.-Y. (2010). Electrochemical deposition and superhydrophobic behavior of ZnO nanorod arrays, *Thin Solid Films*, **518**, pp. 4884–4889.

401. He, G., and Wang K. (2011). The super hydrophobicity of ZnO nanorods fabricated by electrochemical deposition method, *Appl. Surf. Sci.*, **257**, pp. 6590–6594.

402. Hao, Y., Soolaman, D. M., and Yu, H.-Z. (2013). Controlled wetting on electrodeposited oxide thin films: From hydrophilic to superhydrophobic, *J. Phys. Chem. C*, **117**, pp. 7736–7743.

403. Tesler, A. B., Kim, P., Kolle, S., Howell, C., Ahanotu, O., and Aizenberg, J. (2015). Extremely durable biofouling-resistant metallic surfaces based on electrodeposited nanoporous tungstite films on steel, *Nat. Commun.*, **6**, p. 8649.

404. Wang, S., Feng, X., Yao, J., and Jiang, L. (2006). Controlling wettability and photochromism in a dual-responsive tungsten oxide film, *Angew. Chem. Int. Ed.*, **45**, pp. 1264–1267.

405. Cao, L., Liu, J., Xu, S., Xia, Y., Huang, W., and Li, Z. (2013). Inherent superhydrophobicity of Sn/SnOx films prepared by surface self-passivation of electrodeposited porous dendritic Sn, *Mater. Res. Bull.*, **48**, pp. 4804–4810.

406. Zhao, G., Zhang, Y., Lei, Y., Lv, B., Gao, J., Zhang, Y., and Li, D. (2010). Fabrication and electrochemical treatment application of a novel lead oxide anode with superhydrophobic surface, high oxygen evolution potential, and oxidation capability, *Environ. Sci. Technol.*, **44**, pp. 1754–1759.

407. Lei, Y., Zhao, G., Zhang, Y., Liu, M., Liu, L., Lv, B., and Gao, J. (2010). Highly efficient and mild electrochemical incineration: Mechanism and kinetic process of refractory aromatic hydrocarbon pollutants on superhydrophobic Pb_2O anode, *Environ. Sci. Technol.*, **44**, pp. 7921–7927.

408. Liu, Q., Chen, D., and Kang, Z. (2015). One-step electrodeposition process to fabricate corrosion-resistant superhydrophobic surface on magnesium alloy, *ACS Appl. Mater. Interfaces*, **7**, pp. 1859–1867.

409. Liu, Q., and Kang, Z. (2014). One-step electrodeposition process to fabricate superhydrophobic surface with improved anticorrosion property on magnesium alloy, *Mater. Lett.*, **137**, pp. 210–213.
410. Zhang, B., Li, Y., and Hou, B. (2015). One-step electrodeposition fabrication of a superhydrophobic surface on an aluminum substrate with enhanced self-cleaning and anticorrosion properties, *RSC Adv.*, **5**, pp. 100000–100010.
411. Zhang, B., Zhao, X., Li, Y., and Hou, B. (2016). Fabrication of durable anticorrosion superhydrophobic surfaces on aluminum substrates via a facile one-step electrodeposition approach, *RSC Adv.*, **6**, pp. 35455–35465.
412. Liu, C., Su, F., Liang, J., and Huang, P. (2014). Facile fabrication of superhydrophobic cerium coating with micro-nano flower-like structure and excellent corrosion resistance, *Surf. Coat. Technol.*, **258**, pp. 580–586.
413. Liu, Y., Li, S., Zhang, J., Wang, Y., Han, Z., and Ren, L. (2014). Fabrication of biomimetic superhydrophobic surface with controlled adhesion by electrodeposition, *Chem. Eng. J.*, **248**, pp. 440–447.
414. Liu, Y., Li, S., Zhang, J., Liu, J., Han, Z., and Ren, L. (2015). Corrosion inhibition of biomimetic super-hydrophobic electrodeposition coatings on copper substrate, *Corros. Sci.*, **94**, pp. 190–196.
415. Pedraza, F., Mahadik, S. A., and Bouchaud, B. (2015). Synthesis of ceria based superhydrophobic coating on $Ni_{20}Cr$ substrate via cathodic electrodeposition, *Phys. Chem. Chem. Phys.*, **17**, pp. 31750–31757.
416. Zhao, M., Wang, X., Song, H., Li, J., He, G., Gui, Y., and Feng, W. (2015). Fabrication of a superhydrophobic phosphate/fatty-acid salt compound coating on magnesium alloy, *ECS Electrochem. Lett.*, **4**, pp. C19–C21.
417. Chen, Z., Hao, L., and Chen, C. (2012). A fast electrodeposition method for fabrication of lanthanum superhydrophobic surface with hierarchical micro-nanostructures, *Colloids Surf. A*, **401**, pp. 1–7.
418. Wu, L.-K., Hu, J.-M., and Zhang, J.-Q. (2013). One step sol–gel electrochemistry for the fabrication of superhydrophobic surfaces, *J. Mater. Chem. A*, **1**, pp. 14471–14475.
419. Wu, L.-K., Hu, J.-M., Zhang, J.-Q., and Cao, C.-N. (2013). Superhydrophobic surface constructed on electrodeposited sol–gel silica film, *Electrochem. Commun.*, **26**, pp. 85–88.
420. Zhang, X.-F., Chen, R.-J., and Hu, J.-M. (2016). Superhydrophobic surface constructed on electrodeposited silica films by two-step method for corrosion protection of mild steel, *Corros. Sci.*, **104**, pp. 336–343.

421. Zhang, X.-F., Chen, R.-J., Liu, Y.-H., and Hu, J.-M. (2016). Electrochemically generated sol–gel films as inhibitor containers of superhydrophobic surfaces for the active corrosion protection of metals, *J. Mater. Chem. A*, **4**, pp. 649–656.
422. Yin, Y., Huang, R., Zhang, W., Zhang, M., and Wang, C. (2016). Superhydrophobic–superhydrophilic switchable wettability via TiO_2 photoinduction electrochemical deposition on cellulose substrate, *Chem. Eng. J.*, **289**, pp. 99–105.
423. Safaee, A., Sarkar, D. K., and Farzaneh, M. (2008). Superhydrophobic properties of silver-coated films on copper surface by galvanic exchange reaction, *Appl. Surf. Sci.*, **254**, pp. 2493–2498.
424. Gu, C., Ren, H., Tu, J., and Zhang, T.-Y. (2009). Micro/nanobinary structure of silver films on copper alloys with stable water-repellent property under dynamic conditions, *Langmuir*, **25**, pp. 12299–12307.
425. Xu, X., Zhang, Z., and Yang, J. (2010). Fabrication of biomimetic superhydrophobic surface on engineering materials by a simple electroless galvanic deposition method, *Langmuir*, **26**, pp. 3654–3658.
426. Sarkar, D. K., and Paynter, R. W. (2010). One-step deposition process to obtain nanostructured superhydrophobic thin films by galvanic exchange reactions, *J. Adhes. Sci. Technol.*, **24**, pp. 1181–1189.
427. Larmour, I. A., Bell, S. E. J., and Saunders, G. C. (2007). Remarkably simple fabrication of superhydrophobic surfaces using electroless galvanic deposition, *Angew. Chem. Int. Ed.*, **46**, pp. 1710–1712.
428. Xu, P., Wang, F., Yang, C., Ou, J., Li, W., and Amirfazli, A. (2016). Reversible transition between superhydrophobicity and superhydrophilicity on a silver surface, *Surf. Coat. Technol.*, **294**, pp. 47–53.
429. Guo, J., Yu, S., Li, J., and Guo, Z. (2015). Fabrication of functional superhydrophobic engineering materials via an extremely rapid and simple route, *Chem. Commun.*, **51**, pp. 649–6495.
430. Karthik, N., and Sethuraman, M. G. (2015). Transformation of hydrophobic surface into superhydrophobic surface by interfacial flower like silver films, *Surf. Interface Anal.*, **47**, pp. 423–428.
431. Gao, Y., Yang, X., Men, Y., Wang, Y., Xiang, Q., Yue, F., Ding, X., Li, J., Yang, Z., and Wang, Q. (2015). Superhydrophobic inorganic–organic composite coatings with hybrid micro-nano binary structure on copper, *Polym. Adv. Technol.*, **26**, pp. 1320–1325.
432. Rangel, T. C., Michels, A. F., Horowitz, F., and Weibel, D. E. (2015). Superomniphobic and easily repairable coatings on copper substrates

based on simple immersion or spray processes, *Langmuir*, **31**, pp. 3465–3472.

433. Li, P., Chen, X., Yang, G., Yu, L., and Zhang, P. (2014). Fabrication of a superhydrophobic etched copper–silver/stearic acid composite coating and evaluation of its friction-reducing and anticorrosion abilities, *Mater. Exp.*, **4**, pp. 309–316.

434. Li, P., Chen, X., Yang, G., Yu, L., and Zhang, P. (2014). Preparation of silver-cuprous oxide/stearic acid composite coating with superhydrophobicity on copper substrate and evaluation of its friction-reducing and anticorrosion abilities, *Appl. Surf. Sci.*, **289**, pp. 21–26.

435. He, Z., He, J., and Zhang, Z. (2015). Selective growth of metallic nanostructures on microstructured copper substrate in solution, *CrystEngComm*, **17**, pp. 7262–7269.

436. Cheung, M., Lee, W. W. Y., McCracken, J. N., Larmour, I. A., Brennan, S., and Bell, S. E. J. (2016). Raman analysis of dilute aqueous samples by localized evaporation of sub-mL droplets on the tips of superhydrophobic copper wires, *Anal. Chem.*, **88**, pp. 4541–4547.

437. Sang, Y. C., Albadarin, A. B., Al-Muhtaseb, A. H., Mangwandi, C., McCracken, J. N., Bell, S. E. J., and Walker, G. M. (2015). Properties of super-hydrophobic copper and stainless steel meshes: Applications in controllable water permeation and organic solvents/water separation, *Appl. Surf. Sci.*, **335**, pp. 107–114.

438. Cao, Z., Xiao, D., Kang, L., Wang, Z., Zhang, S., Ma, Y., Fu, H., and Yao, J. (2008). Superhydrophobic pure silver surface with flower-like structures by a facile galvanic exchange reaction with [Ag(NH$_3$)$_2$]OH, *Chem. Commun.*, **23**, pp. 2692–2694.

439. Sarkar, D. K., and Saleema, N. (2010). One-step fabrication process of superhydrophobic green coatings, *Surf. Coat. Technol.*, **204**, pp. 2483–2486.

440. Gu, C. D., Xu, X. J., and Tu, J. P. (2010). Fabrication and wettability of nanoporous silver film on copper from choline chloride-based deep eutectic solvents, *J. Phys. Chem. C*, **114**, pp. 13614–13619.

441. Muench, F., Juretzka, B., Narayan, S., Radetinac, A., Flege, S., Schaefer, S., Stark, R. W., and Ensinger, W. (2015). Nano- and microstructured silver films synthesised by halide-assisted electroless plating, *New J. Chem.*, **39**, pp. 6803–6812.

442. Wu, Y., Hang, T., Wang, N., Yu, Z., and Li, M. (2013). Highly durable non-sticky silver film with a microball-nanosheet hierarchical structure prepared by chemical deposition, *Chem. Commun.*, **49**, pp. 10391–10393.

443. Li, K., Zeng, X., Li, H., and Lai, X. (2015). A study on the fabrication of superhydrophobic iron surfaces by chemical etching and galvanic replacement methods and their anti-icing properties, *Appl. Surf. Sci.*, **346**, pp. 458–463.
444. Yu, Z., Song, W., Chen, L., Park, Y., Zhao, B., Cong, Q., and Jung, Y. M. (2015). Simple immersion to prepare a Zn/Ag biomimetic superhydrophobic surface and exploring its applications on SERS, *Colloids Surf. A*, **467**, pp. 224–232.
445. Xu, X., Zhang, Z., Guo, F., Yang, J., Zhu, X., Zhou, X., and Xue, Q. (2012). Superamphiphobic self-assembled monolayer of thiol on the structured Zn surface, *Colloids Surf. A*, **396**, pp. 90–95.
446. Sun, Y., and Qiao, R. (2008). Facile tuning of superhydrophobic states with Ag nanoplates, *Nano Res.*, **1**, pp. 292–302.
447. Ou, J., Shi, Q., Chen, Y., Wang, F., Xue, M., and Li, W. (2015). Superhydrophobic surfaces on diverse metals based on ultrafast sequential deposition of silver and stearic acid, *Appl. Surf. Sci.*, **326**, pp. 139–144.
448. Chen, C.-Y., and Wong, C.-P. (2014). Shape-diversified silver nanostructures uniformly covered on aluminium micro-powders as effective SERS substrates, *Nanoscale*, **6**, pp. 811–816.
449. Guo, W., and Liu, W. (2010). Formation mechanism of robust silver nanoparticle film with superhydrophobicity, *Appl. Phys. Lett.*, **97**, pp. 243701/1–243701/3.
450. Shi, F., Song, Y., Niu, J., Xia, X., Wang, Z., and Zhang, X. (2006). Facile method to fabricate a large-scale superhydrophobic surface by galvanic cell reaction, *Chem. Mater.*, **18**, pp. 1365–1368.
451. Cheng, Y., Lu, S., Xu, W., Wen, H., and Wang, J. (2015). Fabrication of superhydrophobic Au–Zn alloy surface on a zinc substrate for roll-down, self-cleaning and anti-corrosion properties, *J. Mater. Chem. A*, **3**, pp. 16774–16784.
452. Lim, H. S., Lee, S. G., Lee, D. H., Lee, D. Y., Lee, S., and Cho, K. (2008). Superhydrophobic to superhydrophilic wetting transition with programmable ion-pairing interaction, *Adv. Mater.*, **20**, pp. 4438–4441.
453. Wang, C.-H., Song, Y.-Y., Zhao, J.-W., and Xia, X.-H. (2006). Semiconductor supported biomimetic superhydrophobic gold surfaces by the galvanic exchange reaction, *Surf. Sci.*, **600**, pp. 38–42.
454. Zhang, N., Lu, S., Xu, W., and Zhang, Y. (2014). Controlled growth of CuO–Cu$_3$Pt/Cu micro-nano binary architectures on copper substrate and its superhydrophobic behavior, *New J. Chem.*, **38**, pp. 4534–4540.

455. Ning, T., Xu, W., and Lu, S. (2011). Fabrication of superhydrophobic surfaces on zinc substrates and their application as effective corrosion barriers, *Appl. Surf. Sci.*, **258**, pp. 1359–1365.
456. Shi, Y., Yang, W., Bai, J., Feng, X., and Wang, Y. (2014). Fabrication of flower-like copper film with reversible superhydrophobicity-superhydrophilicity and anticorrosion properties, *Surf. Coat. Technol.*, **253**, pp. 148–153.
457. Qiu, R., Zhang, D., and Wang, P. (2013). Superhydrophobic-carbon fibre growth on a zinc surface for corrosion inhibition, *Corros. Sci.*, **66**, pp. 350–359.
458. Cheng, Y., Lu, S., and Xu, W. (2015). Controllable wettability of micro- and nano-dendritic structures formed on aluminum substrates, *New J. Chem.*, **39**, pp. 6602–6610.
459. Karthik, N., and Sethuraman, M. G. (2015). Fabrication of micro-nanocomposite coatings with lotus leaf like texture by combining electroless and candle soot depositions, *New J. Chem.*, **39**, pp. 3337–3340.
460. Yan, W., Liu, H., Chen, T., Sun, Q., and Zhu, W. (2016). A fast and low-cost method to fabricate large-area super-hydrophobic surface on steel substrate with anti-corrosion and anti-icing properties *J. Vac. Sci. Technol. A*, **34**, pp. 041401/1–041401/10.
461. Wang, J., Lu, S., Xu, W., and Zhang, Y. (2014). Synthesis of tin superhydrophobic surfaces on zinc substrates, *RSC Adv.*, **4**, pp. 39197–39203.
462. Wan, B., Ou, J., Lv, D., Xue, M., Wang, F., and Wu, H. (2016). Superhydrophobic ceria on aluminum and its corrosion resistance, *Surf. Interface Anal.*, **48**, pp. 173–178.
463. Su, C., Lu, Z., Zhao, H., Yang, H., and Chen, R. (2015). Photoinduced switchable wettability of bismuth coating with hierarchical dendritic structure between superhydrophobicity and superhydrophilicity, *Appl. Surf. Sci.*, **353**, pp. 735–743.
464. Cao, L., Lu, X., Pu, F., Yin, X., Xia, Y., Huang, W., and Li, Z. (2014). Facile fabrication of superhydrophobic Bi/Bi$_2$O$_3$ surfaces with hierarchical micro-nanostructures by electroless deposition or electrodeposition, *Appl. Surf. Sci.*, **288**, pp. 558–563.
465. Zeng, T., Liao, J., Li, H., Feng, K., and Li, L. (2015). An array of leaf-like Co$_3$Ni microstructures with ferromagnetic properties, superhydrophobic properties and high catalytic performance in the hydrolysis of ammonia borane, *RSC Adv.*, **5**, pp. 105307–105312.
466. Ranjith, K. S., Geethu, R., Vijayakumar, K. P., and Rajendrakumar, R. T. (2014). Control of interconnected ZnO nanowires to vertically

aligned ZnO nanorod arrays by tailoring the underlying spray deposited ZnO seed layer, *Mater. Res. Bull.*, **60**, pp. 584–588.
467. Li, Y., Wang, J., Kong, Y., Zhou, J., Wu, J., Wang, G., Bi, H., Wu, X., Qin, W., and Li, Q. (2016). Micro/nano hierarchical peony-like Al doped ZnO superhydrophobic film: The guiding effect of (100) preferred seed layer, *Sci. Rep.*, **6**, p. 19187.
468. Wang, J., Li, Y., Kong, Y., Zhou, J., Wu, J., Wu, X., Qin, W., Jiao, Z., and Jiang, L. (2015). Non-fluorinated superhydrophobic and micro/nano hierarchical Al doped ZnO film: The effect of Al doping on morphological and hydrophobic properties, *RSC Adv.*, **5**, pp. 81024–81029.
469. Laurenti, M., Cauda, V., Gazia, R., Fontana, M., Rivera, V. F., Bianco, S., and Canavese, G. (2013). Wettability control on ZnO nanowires driven by seed layer properties, *Eur. J. Inorg. Chem.*, **2013**, pp. 2520–2527.
470. Zhang, Y., Fang, F., Wang, C., Wang, L., Wang, X., Chu, X., Li, J., Fang, X., Wei, Z., and Wang, X. (2014). Hydrophobic modification of ZnO nanostructures surface using silane coupling agent, *Polym. Compos.*, **35**, pp. 1204–1211.
471. Gao, D., and Jia, M. (2015). Preparation of hierarchical porous Zn-salt particles and their superhydrophobic performance, *Appl. Surf. Sci.*, **359**, pp. 89–97.
472. Li, J., Yang, Y., Zha, F., and Lei, Z. (2012). Facile fabrication of superhydrophobic ZnO surfaces from high to low water adhesion, *Mater. Lett.*, **75**, pp. 71–73.
473. Rezayia, T., and Entezari, M. H. (2016). Wettability properties vary with different morphologies of ZnO nanoparticles deposited on glass and modified by stearic acid, *New J. Chem.*, **40**, pp. 2582–2591.
474. Zheng, L., Li, Z., Bourdo, S., Saini, V., Ryerson, C., and Biris, A. S. (2011). Hierarchical ZnO structure with superhydrophobicity and high adhesion, *ChemPhysChem*, **12**, pp. 2412–2414.
475. Kwak, G., Jung, S., and Yong, K. (2011). Multifunctional transparent ZnO nanorod films, *Nanotechnology*, **22**, pp. 115705/1–115705/7.
476. Tian, J., Zhang, Y., Zhu, J., Yang, Z., and Gao, X. (2014). Robust nonsticky superhydrophobicity by the tapering of aligned ZnO nanorods, *ChemPhysChem*, **15**, pp. 858–861.
477. Zhang, J., and Zhang, J. (2013). A facile method for preparing a non-adhesive superhydrophobic ZnO nanorod surface, *Mater. Lett.*, **93**, pp. 386–389.
478. Sutha, S., Vanithakumari, S. C., George, R. P., Mudali, U. K., Raj, B., and Ravi, K. R. (2015). Studies on the influence of surface morphology

of ZnO nail beds on easy roll off of water droplets, *Appl. Surf. Sci.*, **347**, pp. 839–848.

479. Wang, G., Zeng, Z., Chen, J., Xu, M., Zhu, J., Liu, S., Ren, T., and Xue, Q. (2016). Ultra low water adhesive metal surface for enhanced corrosion protection, *RSC Adv.*, **6**, pp. 40641–40649.

480. Li, B.-J., Huang, L.-J., Ren, N.-F., Kong, X., Cai, Y.-L., and Zhang, J.-L. (2016). Superhydrophobic and anti-reflective ZnO nanorod-coated FTO transparent conductive thin films prepared by a three-step method, *J. Alloys Compounds*, **674**, pp. 368–375.

481. Chen, Y., Yang, G., and Jing, Z. (2016). Synthesis and characterization of superhydrophobic CeO_2/ZnO nanotube arrays with low adhesive force, *Mater. Lett.*, **176**, pp. 290–293.

482. Lee, D. J., Kim, H. M., Song, Y. S., and Youn, J. R. (2012). Water droplet bouncing and superhydrophobicity induced by multiscale hierarchical nanostructures, *ACS Nano*, **6**, pp. 7656–7664.

483. Ranjbar, M., Taher, M. A., and Sam, A. (2016). Facile single-step synthesis of SiO_2-coated ZnO nanorod as hydrophobic layer by hydrothermal method, *J. Cluster Sci.*, **27**, pp. 105–114.

484. Xue, M., Xu, T., Xie, X., Ou, J., Wang, F., and Li, W. (2015). Formation, transformation and superhydrophobicity of compound surfactant-assisted aligned ZnO nanoplatelets, *Appl. Surf. Sci.*, **355**, pp. 1063–1068.

485. Wang, T., Chang, L., Zhuang, L., Yang, S., Jia, Y., and Wong, C. (2015). A hierarchical and superhydrophobic ZnO/C surface derived from a rice-leaf template, *Monatsh. Chem.*, **145**, pp. 65–69.

486. Dai, S., Zhang, D., Shi, Q., Han, X., Wang, S., and Du, Z. (2013). Biomimetic fabrication and tunable wetting properties of three-dimensional hierarchical ZnO structures by combining soft lithography templated with lotus leaf and hydrothermal treatments, *CrystEngComm*, **15**, pp. 5417–5424.

487. Palamà, I. E., D'Amone, S., Biasiucci, M., Gigli, G., and Cortese, B. (2014). Bioinspired design of a photoresponsive superhydrophobic/oleophilic surface with underwater superoleophobic efficacy, *J. Mater. Chem. A*, **2**, pp. 17666–17675.

488. Hao, X., Wang, L., Lv, D., Wang, Q., Li, L., He, N., and Lu, B. (2015). Fabrication of hierarchical structures for stable superhydrophobicity on metallic planar and cylindrical inner surfaces, *Appl. Surf. Sci.*, **325**, pp. 151–159.

489. Liu, Y., Das, A., Xu, S., Lin, Z., Xu, C., Wang, Z. L., Rohatgi, A., and Wong, C. P. (2012). Hybridizing ZnO nanowires with micropyramid silicon wafers as superhydrophobic high-efficiency solar cells, *Adv. Energy Mater.*, **2**, pp. 47–51.

490. Lee, J., and Yong, K. (2015). Combining the lotus leaf effect with artificial photosynthesis: Regeneration of underwater superhydrophobicity of hierarchical ZnO/Si surface by solar water splitting, *NPG Asia Mater.*, **7**, p. e201.

491. Tao, Q., Li, S., Ma, C., Liu, K., and Zhang, Q.-Y. (2015). A highly sensitive and recyclable SERS substrate based on Ag-nanoparticle-decorated ZnO nanoflowers in ordered arrays, *Dalton Trans.*, **44**, pp. 3447–3453.

492. Hao, X., Lv, D., Wang, L., Luo, Y., Li, L., and He, N. (2017). Controllable growth and characterization of zinc oxide nanopillars arrays by a facile hydrothermal, *Microelectron. Eng.*, in press.

493. Sun, H., Luo, M., Weng, W., Cheng, K., Du, P., Shen, G., and Han, G. (2008). Position and density control in hydrothermal growth of ZnO nanorod arrays through pre-formed micro/nanodots, *Nanotechnology*, **19**, pp. 395602/1–395602/7.

494. Yu, H., Liu, J., Fan, X., Yan, W., Han, L., Han, J., Zhang, X., Hong, T., and Liu, Z. (2016). Bionic micro-nano-bump-structures with a good self-cleaning property: The growth of ZnO nanoarrays modified by polystyrene spheres, *Mater. Chem. Phys.*, **170**, pp. 52–61.

495. Shin, Y.-M., Lee, S.-K., Lee, J.-Y., Kim, J.-H., Park, J.-H., and Ji, C.-H. (2013). Microfabricated environmental barrier using ZnO nanowire on metal mesh, *J. Micromech. Microeng.*, **23**, pp. 127001/1–127001/6.

496. Ottone, C., Lamberti, A., Fontana, M., and Cauda, V. (2014). Wetting behavior of Hierarchical oxide nanostructures: TiO$_2$ nanotubes from anodic oxidation decorated with ZnO nanostructures, *J. Electrochem. Soc.*, **161**, pp. D484–D488.

497. Li, H., and Yu, S. (2016). Facile fabrication of micro–nano-rod structures for inducing a superamphiphobic property on steel surface, *Appl. Phys. A*, **122**, p. 30.

498. Guo, Z., Chen, X., Li, J., Liu, J.-H., and Huang, X.-J. (2011). ZnO/CuO hetero-hierarchical nanotrees array: Hydrothermal preparation and self-cleaning properties, *Langmuir*, **27**, pp. 6193–6200.

499. Lin, X., Lu, F., Chen, Y., Liu, N., Cao, Y., Xu, L., Wei, Y., and Feng, L. (2015). One-step breaking and separating emulsion by tungsten oxide coated mesh, *ACS Appl. Mater. Interfaces*, **7**, pp. 8108–8113.

500. Fan, Y., Chen, Z., Liang, J., Wang, Y., and Chen, H. (2014). Preparation of superhydrophobic films on copper substrate for corrosion protection, *Surf. Coat. Technol.*, **244**, pp. 1–8.

501. Li, M., Su, Y., Hu, J., Yao, L., Wei, H., Yang, Z., and Zhang, Y. (2016). Hierarchically porous micro/nanostructured copper surfaces with enhanced antireflection and hydrophobicity, *Appl. Surf. Sci.*, **361**, pp. 11–17.

502. Zhao, M., Shang, F., Song, Y., Wang, F., Zhou, Z., Lv, J., Zi, Z., Wei, Y., Chen, X., He, G., Zhang, M., Song, X., and Sun, Z. (2015). Surface morphology, composition and wettability Cu_2O/CuO composite thin films prepared by a facile hydrothermal method, *Appl. Phys. A*, **118**, pp. 901–906.

503. Wang, T., Liu, G., and Kong, J. (2015). Preparation of wood-like structured copper with superhydrophobic properties, *Sci. Rep.*, **5**, p. 18328.

504. Cao, H., Zheng, H., Yin, J., Lu, Y., Wu, S., Wu, X., and Li, B. (2010). $Mg(OH)_2$ complex nanostructures with superhydrophobicity and flame retardant effects, *J. Phys. Chem. C*, **114**, pp. 17362–17368.

505. Shi, F., Chen, X., Wang, L., Niu, J., Yu, J., Wang, Z., and Zhang, X. (2005). Roselike microstructures formed by direct in situ hydrothermal synthesis: From superhydrophilicity to superhydrophobicity, *Chem. Mater.*, **17**, pp. 6177–6180.

506. Wang, Z., Tian, Y., Fan, H., Gong, J., Yang, S., Ma, J., and Xu, J. (2014). Facile seed-assisted hydrothermal fabrication of γ-AlOOH nanoflake films with superhydrophobicity, *New J. Chem.*, **38**, pp. 1321–1327.

507. Song, K. M., Ahn, S. H., and Cho, Y.-S. (2015). Fabrication of a hierarchical aluminum oxide surface with micro/nanostructures via a single process and its application as a superhydrophobic surface: Simple sintering method with an aluminum microsized powder, *Surf. Coat. Technol.*, **282**, pp. 68–77.

508. Liu, L., Feng, X., and Guo, M. (2013). Eco-friendly fabrication of superhydrophobic bayerite array on Al foil via an etching and growth process, *J. Phys. Chem. C*, **117**, pp. 25519–25525.

509. Liu, L., Yang, L.-Q., Liang, H.-W., Cong, H.-P., Jiang, J., and Yu, S.-H. (2013). Bio-inspired fabrication of hierarchical FeOOH nanostructure array films at the air–water interface, their hydrophobicity and application for water treatment, *ACS Nano*, **7**, pp. 1368–1378.

510. Wang, S., Wang, C., Liu, C., Zhang, M., Ma, H., and Li, J. (2012). Fabrication of superhydrophobic spherical-like α-FeOOH films on

the wood surface by a hydrothermal method, *Colloids Surf. A*, **403**, pp. 29–34.

511. Cao, H., Zheng, H., Liu, K., and Warner, J. H. (2010). Bioinspired peony-like beta-Ni(OH)$_2$ nanostructures with enhanced electrochemical activity and superhydrophobicity, *ChemPhysChem*, **11**, pp. 489–494.

512. Abellán, G., Carrasco, J. A., Coronado, E., Prieto-Ruiz, J. P., and Prima-García, H. (2014). In-situ growth of ultrathin films of NiFe-LDHs: Towards a hierarchical synthesis of bamboo-like carbon nanotubes, *Adv. Mater. Interfaces*, **1**, pp. 1400184/1–1400184/10.

513. Zhang, X., Guo, Y., Liu, Y., Yang, X., Pan, J., and Zhang, P. (2013). Facile fabrication of superhydrophobic surface with nanowire structures on nickel foil, *Appl. Surf. Sci.*, **287**, pp. 299–303.

514. Yang, M., Neupane, S., Wang, X., He, J., Li, W., and Pala, N. (2013). Multiple step growth of single crystalline rutile nanorods with the assistance of self-assembled monolayer for dye sensitized solar cells, *ACS Appl. Mater. Interfaces*, **5**, pp. 9809–9815.

515. Choi, H.-J., Shin, J.-H., Choo, S., Kim, J., and Lee, H. (2013). A tunable method for nonwetting surfaces based on nanoimprint lithography and hydrothermal growth, *J. Mater. Chem. A*, **1**, pp. 8417–8424.

516. Liu, M., Qing, Y., Wu, Y., Liang, J., and Luo, S. (2015). Facile fabrication of superhydrophobic surfaces on wood substrates via a one-step hydrothermal process, *Appl. Surf. Sci.*, **330**, pp. 332–338.

517. Shen, Y., Tao, J., Tao, H., Chen, S., Pan, L., and Wang, T. (2015). Nanostructures in superhydrophobic Ti$_6$Al$_4$V hierarchical surfaces control wetting state transitions, *Soft Matter*, **11**, pp. 3806–3811.

518. Shen, Y., Tao, J., Tao, H., Chen, S., Pan, L., and Wang, T. (2015). Superhydrophobic Ti$_6$Al$_4$V surfaces with regular array patterns for anti-icing applications, *RSC Adv.*, **5**, pp. 32813–32818.

519. Shen, Y., Tao, J., Tao, H., Chen, S., Pan, L., and Wang, T. (2015). Anti-icing potential of superhydrophobic Ti$_6$Al$_4$V surfaces: Ice nucleation and growth, *Langmuir*, **31**, pp. 10799–10806.

520. Tokudome, Y., Okada, K., Nakahira, A., and Takahashi, M. (2014). Switchable and reversible water adhesion on superhydrophobic titanate nanostructures fabricated on soft substrates: Photopatternable wettability and thermomodulatable adhesivity, *J. Mater. Chem. A*, **2**, pp. 58–61.

521. Dong, R., Jiang, S., Li, Z., Chen, Z., Zhang, H., and Jin, C. (2015). Superhydrophobic TiO$_2$ nanorod films with variable morphology grown on different substrates, *Mater. Lett.*, **62**, pp. 151–154.

522. Li, L., Huang, T., Lei, J., He, J., Qu, L., Huang, P., Zhou, W., Li, N., and Pan, F. (2015). Robust biomimetic-structural superhydrophobic surface on aluminum alloy, *ACS Appl. Mater. Interfaces*, **7**, pp. 1449–1457.

523. Wang, D., Guo, Z., Chen, Y., Hao, J., and Liu, W. (2007). In situ hydrothermal synthesis of nanolamellate $CaTiO_3$ with controllable structures and wettability, *Inorg. Chem.*, **46**, pp. 7707–7709.

524. Yao, Q., Jin, C., Zheng, H., Ma, Z., and Sun Q. (2015). Superhydrophobicity, microwave absorbing property of $NiFe_2O_4$/wood hybrids under harsh conditions, *J. Nanomater.*, **2015**, pp. 761286/1–761286/8.

525. Chen, Y., Li, F., Li, T., and Cao, W. (2016). Shape-controlled hydrothermal synthesis of superhydrophobic and superoleophilic $BaMnF_4$ micro/nanostructures, *CrystEngComm*, **18**, pp. 3585–3593.

526. Gu, K.-C., Chen, B.-S., Wang, X.-M., Wang, J., Fang, J.-H., Wu, J., and Huang, L.-C. (2014). Preparation, friction and wear properties of hydrophobic lanthanum borate nanorods in rapeseed oil, *Trans. Nonferrous Met. Soc. China*, **24**, pp. 3578–3584.

527. Yang, C., Zhang, L., Wang, Z., Li, T., Li, F., and Cao, W. (2016). Nanostructured $NaLa(MoO_4)_2$ and Eu^{3+}-doped $NaLa(MoO_4)_2$: Synthesis, characterizations, photoluminscence and superhydrophobic properties, *Mater. Sci. Eng. B*, **207**, pp. 39–46.

528. Chen, H., Zou, R., Wang, N., Chen, H., Zhang, Z., Sun, Y., Yu, L., Tian, Q., Chen, Z., and Hu, J. (2011). Morphology-selective synthesis and wettability properties of well-aligned $Cu_{2-x}Se$ nanostructures on a copper substrate, *J. Mater. Chem.*, **21**, pp. 3053–3059.

529. Zhang, Q., Yang, J., Wang, C.-F., Chen, Q.-L., and Chen, S. (2013). Facile fabrication of fluorescent-superhydrophobic bifunctional ligand-free quantum dots, *Colloid Polym. Sci.*, **291**, pp. 717–723.

530. Lian, G., Zhang, X., Tan, M., Zhang, S., Cui, D., and Wang, Q. (2011). Facile synthesis of 3D boron nitride nanoflowers composed of vertically aligned nanoflakes and fabrication of graphene-like BN by exfoliation, *J. Mater. Chem.*, **21**, pp. 9201–9207.

531. Zhou, S., Hao, G., Zhou, X., Jiang, W., Wang, T., Zhang, N., and Yu, N. (2016). One-pot synthesis of robust suoperhydrophobic, functionalized graphene/polyurethane sponge for effective continuous oil-water separation, *Chem. Eng. J.*, **302**, pp. 155–162.

532. Luo, H., Ma, J., Wang, P., Bai, J., and Jing, G. (2015). Two-step wetting transition on ZnO nanorod arrays, *Appl. Surf. Sci.*, **347**, pp. 868–874.

533. Wang, L., Zhang, X., Li, B., Sun, P., Yang, J., Xu, H., and Liu, Y. (2011). Superhydrophobic and ultraviolet-blocking cotton textiles, *ACS Appl. Mater. Interfaces*, **3**, pp. 1277-1281.
534. Athauda, T. J., Hari, P., and Ozer, R. R. (2013). Tuning physical and optical properties of ZnO nanowire arrays grown on cotton fibers, *ACS Appl. Mater. Interfaces*, **5**, pp. 6237-6246.
535. Jia, W., Jia, B., Wu, X., and Qu, F. (2012). Self assembly of shape-controlled ZnS nanostructures with novel yellow light photoluminescence and excellent hydrophobic properties, *CrystEngComm*, **14**, pp. 7759-7763.
536. Gupta, A., Mondal, K., Sharma, A., and Bhattacharya, S. (2015). Superhydrophobic polymethylsilsesquioxane pinned one dimensional ZnO nanostructures for water remediation through photo-catalysis, *RSC Adv.*, **5**, pp. 45897-45907.
537. Lee, M., and Yong, K. (2012). Highly efficient visible light photocatalysis of novel CuS/ZnO heterostructure nanowire arrays, *Nanotechnology*, **23**, pp. 194014/1-194010/6.
538. Cao, H., Zheng, H., Liu, K., and Fu, R. (2010). Single-crystalline semiconductor In(OH)$_3$ nanocubes with bifunctions: Superhydrophobicity and photocatalytic activity, *Cryst. Growth Des.*, **10**, pp. 597-601.
539. Yang, C., Huang, Y., Li, F., and Li, T. (2016). One-step synthesis of Bi$_2$WO$_6$/TiO$_2$ heterojunctions with enhanced photocatalytic and superhydrophobic property via hydrothermal method, *J. Mater. Sci.*, **51**, pp. 1032-1042.
540. Yang, C., Huang, Y., Li, T., and Li, F. (2015). Bi$_2$WO$_6$ nanosheets synthesized by a hydrothermal method: Photocatalytic activity driven by visible light and the superhydrophobic property with water adhesion, *Eur. J. Inorg. Chem.*, **2015**, pp. 2560-2564.
541. Yang, C., Yang, X, Li, F., Li, T., and Cao, W. (2016). Controlled synthesis of hierarchical flower-like Sb$_2$WO$_6$ microspheres: Photocatalytic and superhydrophobic property, *J. Ind. Eng. Chem.*, **39**, pp. 93-100.
542. Yang, X., Zuo, W., Li, F., and Li, T. (2015). Surfactant-free and controlled synthesis of hexagonal CeVO$_4$ nanoplates: Photocatalytic activity and superhydrophobic property, *ChemistryOpen*, **4**, pp. 288-294.
543. Guo, Z., Chen, B., Zhang, M., Mu, J., Shao, C., and Liu, Y. (2010). Zinc phthalocyanine hierarchical nanostructure with hollow interior space: Solvent-thermal synthesis and high visible photocatalytic property, *J. Colloid Interface Sci.*, **348**, pp. 37-42.
544. Yada, M., Inoue, Y., Sakamoto, A., Torikai, T., and Watari, T. (2014). Synthesis and controllable wettability of micro- and nanostructured

titanium phosphate thin films formed on titanium plates, *ACS Appl. Mater. Interfaces*, **6**, pp. 7695–7704.

545. Kim, D. H., Park, J.-H., Lee, T. I., and Myoung, J.-M. (2016). Superhydrophobic Al-doped ZnO nanorods-based electrically conductive and self-cleanable antireflecting window layer for thin film solar cell, *Solar Energy Mater. Solar Cells*, **150**, pp. 65–70.

546. Hiralal, P., Chien, C., Lal, N. N., Abeygunasekara, W., Kumar, A., Butt, H., Zhou, H., Unalan, H. E., Baumberg, J. J., and Amaratunga, G. A. J. (2014). Nanowire-based multifunctional antireflection coatings for solar cells, *Nanoscale*, **6**, pp. 14555–14562.

547. Wu, J., Xia, J., Lei, W., and Wang, B.-P. (2010). Electrowetting on ZnO nanowires, *Appl. Phys. A*, **99**, pp. 931–934.

548. Gu, C., Zhang, J., and Tu, J. (2010). A strategy of fast reversible wettability changes of WO_3 surfaces between superhydrophilicity and superhydrophobicity, *J. Colloid Interface Sci.*, **352**, pp. 573–579.

549. Sonia, S., Suresh Kumar, P., Jayram, N. D., Masuda, Y., Devanasen, M., and Lee, C. (2016). Superhydrophobic and H_2S gas sensing properties of CuO nanostructured thin films through successive ionic layered adsorption reaction process, *RSC Adv.*, **6**, pp. 24290–24298.

550. Li, R., Han, C., and Chen, Q.-W. (2013). A facile synthesis of multifunctional ZnO/Ag sea urchin-like hybrids as highly sensitive substrates for surface-enhanced Raman detection, *RSC Adv.*, **3**, pp. 11715–11722.

551. Kuo, C.-H., Li, W., Song, W., Luo, Z., Poyraz, A. S., Guo, Y., Ma, A. W. K., Sui, S. L., and He, J. (2014). Facile synthesis of Co_3O_4@CNT with high catalytic activity for CO oxidation under moisture-rich conditions, *ACS Appl. Mater. Interfaces*, **6**, pp. 11311–11317.

552. Wang, P., Han, H., Li, J., Fan, X., Ding, H., and Wang, J. (2016). A facile cost-effective method for preparing poinsettia-inspired superhydrophobic ZnO nanoplate surface on Al substrate with corrosion resistance, *Appl. Phys. A*, **122**, p. 53.

553. Shi, Y., Yang, W., Feng, X., Wang, Y., and Yue, G. (2015). Fabrication of superhydrophobic ZnO nanorods surface with corrosion resistance via combining thermal oxidation and surface modification, *Mater. Lett.*, **151**, pp. 24–27.

554. Cho, Y. J., Jang, H., Lee, K.-S., and Kim, D. R. (2015). Direct growth of cerium oxide nanorods on diverse substrates for superhydrophobicity and corrosion resistance, *Appl. Surf. Sci.*, **340**, pp. 96–101.

555. Zhang, K., Wu, J., Chu, P., Ge, Y., Zhao, R., and Li, X. (2015). A novel CVD method for rapid fabrication of superhydrophobic surface

on aluminum alloy coated nanostructured cerium-oxide and its corrosion resistance, *Int. J. Electrochem. Sci.*, **10**, pp. 6257–6272.
556. Zhou, M., Pang, X., Wei, L., and Gao, K. (2015). Insitu grown superhydrophobic Zn–Al layered double hydroxides films on magnesium alloy to improve corrosion properties, *Appl. Surf. Sci.*, **337**, pp. 172–177.
557. Zhang, X., Wu, G., Peng, X., Li, L., Feng, H., Gao, B., Huo, K., and Chu, P. K. (2015). Mitigation of corrosion on magnesium alloy by predesigned surface corrosion, *Sci. Rep.*, **5**, pp. 17399.
558. Gao, R., Liu, Q., Wang, J., Zhang, X., Yang, W., Liu, J., and Liu, L. (2014). Fabrication of fibrous szaibelyite with hierarchical structure superhydrophobic coating on AZ31 magnesium alloy for corrosion protection, *Chem. Eng. J.*, **241**, pp. 352–359.
559. Li, T., Li, Q., Yan, J., and Li, F. (2014). Facile fabrication of corrosion-resistant superhydrophobic and superoleophilic surfaces with $MnWO_4:Dy^{3+}$ microbouquets, *Dalton Trans.*, **43**, pp. 5801–5805.
560. Liu, L., Chen, R., Liu, W., Zhang, Y., Shi, X., and Pan, Q. (2015). Fabrication of superhydrophobic copper sulfide film for corrosion protection of copper, *Surf. Coat. Technol.*, **272**, pp. 221–228.
561. Gao, R., Liu, Q., Wang, J., Liu, J., Yang, W., Gao, Z., and Liu, L. (2014). Construction of superhydrophobic and superoleophilic nickel foam for separation of water and oil mixture, *Appl. Surf. Sci.*, **289**, pp. 417–424.
562. Hoshyargar, F., Mahajan Anuradha, M., Bhosale, S. V., Kyratzis, L., Bhatt, A. I., and O'Mullane, A. P. (2016). Superhydrophobic fabrics for oil-water separation based on the metal organic charge transfer complex CuTCNAQ, *ChemPlusChem*, **81**, pp. 378–383.
563. Dutta, K., and Pramanik, A. (2013). Synthesis of a novel cone-shaped CaAl-layered double hydroxide (LDH): Its potential use as a reversible oil sorbent, *Chem. Commun.*, **49**, pp. 6427–6429.
564. Motlagh, N. V., Birjandi, F. C., Sargolzaei, J., and Shahtahmassebi, N. (2013). Durable, superhydrophobic, superoleophobic and corrosion resistant coating on the stainless steel surface using a scalable method, *Appl. Surf. Sci.*, **283**, pp. 636–647.
565. Motlagh, N. V., Sargolzaei, J., and Shahtahmassebi, N. (2013). Super-liquid-repellent coating on the carbon steel surface, *Surf. Coat. Technol.*, **235**, pp. 241–249.
566. Li, J., Zhao, Z., Zhang, Y., Xiang, B., Tang, X., and She, H. (2016). Facile fabrication of superhydrophobic silica coatings with excellent corrosion resistance and liquid marbles, *J. Sol-Gel Sci. Technol.*, **80**, pp. 208–214.

567. Basu, B. J., Hariprakash, V., Aruna, S. T., Lakshmi, R. V., Manasa, J., and Shruthi, B. S. (2010). Effect of microstructure and surface roughness on the wettability of superhydrophobic sol–gel nanocomposite coatings, *J. Sol-Gel Sci. Technol.*, **56**, pp. 278–286.

568. Li, J., Yan, L., Ouyang, Q., Zha, F., Jing, Z., Li, X., and Lei, Z. (2014). Facile fabrication of translucent superamphiphobic coating on paper to prevent liquid pollution, *Chem. Eng. J.*, **246**, pp. 238–243.

569. Ge, D., Yang, L., Zhang, Y., Rahmawan, Y., and Yang, S. (2014). Transparent and superamphiphobic surfaces from one-step spray coating of stringed silica nanoparticle/sol solutions, *Part. Part. Syst. Charact.*, **31**, pp. 763–770.

570. Rao, A. V., Latthe, S. S., Mahadik, S. A., and Kappenstein, C. (2011). Mechanically stable and corrosion resistant superhydrophobic sol–gel coatings on copper substrate, *Appl. Surf. Sci.*, **257**, pp. 5772–5776.

571. Zhang, G., Wang, D., Gu, Z. Z., and Möhwald, H. (2005). Fabrication of superhydrophobic surfaces from binary colloidal assembly, *Langmuir*, **21**, pp. 9143–9148.

572. Geng, Z., He, J., Xu, L., and Yao, L. (2013). Rational design and elaborate construction of surface nano-structures toward highly antireflective superamphiphobic coatings, *J. Mater. Chem. A*, **1**, pp. 8721–8724.

573. Geng, Z., and He, J. (2014). An effective method to significantly enhance the robustness and adhesion-to-substrate of high transmittance superamphiphobic silica thin films, *J. Mater. Chem. A*, **2**, pp. 16601–16607.

574. Nakajima, A., Saiki, C., Hashimoto, K., and Watanabe, T. (2001). Processing of roughened silica film by coagulated colloidal silica for super-hydrophobic coating, *J. Mater. Sci. Lett.*, **20**, pp. 1975–1977.

575. Hikita, M., Tanaka, K., Nakamura, T., Kajiyama, T., and Takahara, A. (2005). Super-liquid-repellent surfaces prepared by colloidal silica nanoparticles covered with fluoroalkyl groups, *Langmuir*, **21**, pp. 7299–7302.

576. Xu, Q. F., Wang, J. N., and Sanderson, K. D. (2010). Organic–inorganic composite nanocoatings with superhydrophobicity, good transparency, and thermal stability, *ACS Nano*, **4**, pp. 2201–2209.

577. Amigoni, S., Taffin de Givenchy, E., Dufay, M., and Guittard, F. (2009). Covalent layer-by-layer assembled superhydrophobic organic-inorganic hybrid films, *Langmuir*, **25**, pp. 11073–11077.

578. Du, X., and He, J. (2011). A self-templated etching route to surface-rough silica nanoparticles for superhydrophobic coatings, *ACS Appl. Mater. Interfaces*, **3**, pp. 1269–1276.

579. Cao, L., and Gao, D. (2010). Transparent superhydrophobic and highly oleophobic coatings, *Faraday Discuss.*, **146**, pp. 57–65.
580. Soliveri, G., Annunziata, R., Ardizzone, S., Cappelletti, G., and Meroni, D. (2012). Multiscale rough titania films with patterned hydrophobic/oleophobic features, *J. Phys. Chem. C*, **116**, pp. 26405–26413.
581. Li, Q., Wang, Y., Rong, C., Zhang, F., Liu, Y., Chen, L., Wang, Q., and Peng, C. (2016). Facile assembly of graphene and titania on microstructured substrates for superhydrophobic surfaces, *Ceramics Int.*, **42**, pp. 2829–2835.
582. Yang, J., Zhang, Z.-Z., Men X.-H., and Xu, X.-H. (2011). Superoleophobicity of a material made from fluorinated titania nanoparticles, *J. Dispers. Sci. Technol.*, **32**, pp. 485–489.
583. Soliveri, G., Annunziata, R., Ardizzone, S., Cappelletti, G., and Meroni, D. (2012). Multiscale rough titania films with patterned hydrophobic/oleophobic features, *J. Phys. Chem. C*, **116**, pp. 26405–26413.
584. Qi, C., Zheng, Y., Cao, L., Gao, J., and Wan, Y. (2016). Preparation and performance of sol–gel-derived alumina film modified by stearic acid, *J. Sol-Gel Sci. Technol.*, **78**, pp. 641–646.
585. Bao, X.-M., Cui, J.-F., Sun, H.-X., Liang, Wei-D., Zhu, Z.-Q., An, J., Yang, B.-P., La, P.-Q., and Li, A. (2014). Facile preparation of superhydrophobic surfaces based on metal oxide nanoparticles, *Appl. Surf. Sci.*, **303**, pp. 473–480.
586. Qiu, Z., Sun, J., Wang, R., Zhang, Y., and Wu, X. (2016). Magnet-induced fabrication of a superhydrophobic surface on ZK60 magnesium alloy, *Surf. Coat. Technol.*, **286**, pp. 246–250.
587. Bayata, A., Ebrahimi, M., Nourmohammadi, A., and Moshfegh, A.Z. (2015). Wettability properties of PTFE/ZnO nanorods thin film exhibiting UV-resilient superhydrophobicity, *Appl. Surf. Sci.*, **341**, pp. 92–99.
588. Shen, L., Ji, J., and Shen, J. (2008). Silver mirror reaction as an approach to construct superhydrophobic surfaces with high reflectivity, *Langmuir*, **24**, pp. 9962–9965.
589. Jin, C., Li, J., Han, S., Wang, J., Yao, Q., and Sun, Q. (2015). Silver mirror reaction as an approach to construct a durable, robust superhydrophobic surface of bamboo timber with high conductivity, *J. Alloys Compd.*, **635**, pp. 300–306.
590. Yang, J., Zhang, Z., Men, X., Xu, X., and Zhu, X. (2011). A simple approach to fabricate superoleophobic coatings, *New J. Chem.*, **35**, pp. 576–580.

591. Yang, J., Zhang, Z., Men, X., Xu, X., and Zhu, X. (2010). A simple approach to fabricate regenerable superhydrophobic coatings, *Colloids Surf. A*, **367**, pp. 60–64.

592. Li, J., Wu, R., Jing, Z., Yan, L., Zha, F., and Lei, Z. (2015). One-step spray-coating process for the fabrication of colorful superhydrophobic coatings with excellent corrosion resistance, *Langmuir*, **31**, pp. 10702–10707.

593. Hsieh, C.-T., Tzou, D.-Y., Pan, C., and Chen, W.-Y. (2012). Microwave-assisted deposition, scalable coating, and wetting behavior of silver nanowire layers, *Surf. Coat. Technol.*, **207**, pp. 11–18.

594. Li, X., Lee, H. K., Phang, I. Y., Lee, C. K., and Ling, X. Y. (2014). Superhydrophobic-oleophobic Ag nanowire platform: An analyte-concentrating and quantitative aqueous and organic toxin surface-enhanced Raman scattering sensor, *Anal. Chem.*, **86**, pp. 10437–10444.

595. Wang, Z., Ou, J., Wang, Y., Xue, M., Wang, F., Pan, B., Li, C., and Li, W. (2015). Anti-bacterial superhydrophobic silver on diverse substrates based on the mussel-inspired polydopamine. *Surf. Coat. Technol.*, **280**, pp. 378–383.

596. Wang, J., Guo, J., Si, P., Cai, W., Wang, Y., and Wu, G. (2016). Polydopamine-based synthesis of an In(OH)$_3$–PDMS sponge for ammonia detection by switching surface wettability, *RSC Adv.*, **6**, pp. 4329–4334.

597. Zhong, X., Zhao, H., Yang, H., Liu, Y., Yan, G., and Chen, R. (2014). Tunable surface wettability and water adhesion of Sb$_2$S$_3$ micro-/nanorod films, *Appl. Surf. Sci.*, **289**, pp. 425–429.

598. Zhang, Y., Zhang, Q., Wang, C.-F., and Chen, S. (2013). Interfacial self-assembly of Ni$_x$Cd$_{1-x}$S/ODA hybrids with photoluminescent and superhydrophobic performance, *Ind. Eng. Chem. Res.*, **52**, pp. 11590–11596.

599. Sun, Z., Liao, T., Liu, K., Jiang, L., Kim, J. H., and Dou, S. X. (2013). Robust superhydrophobicity of hierarchical ZnO hollow microspheres fabricated by two-step self-assembly, *Nano Res.*, **6**, pp. 726–735.

600. Kakade, B. A. (2013). Chemical control of superhydrophobicity of carbon nanotube surfaces: Droplet pinning and electrowetting behavior, *Nanoscale*, **5**, pp. 7011–7016.

601. Meng, L.-Y., and Park, S.-J. (2012). Effect of growth of graphite nanofibers on superhydrophobic and electrochemical properties of carbon fibers, *Mater. Chem. Phys.*, **132**, pp. 324–329.

602. Wang, Y., Wang, L., Wang, S., Wood, R. J. K., and Xue, Q. (2012). From natural Lotus leaf to highly hard-flexible diamond-like carbon surface with superhydrophobic and good tribological performance, *Surf. Coat. Technol.*, **206**, pp. 2258–2264.
603. Deng, X., Mammen, L., Butt, H.-J., and Vollmer, D. (2012). Candle soot as a template for a transparent robust superamphiphobic coating, *Science*, **335**, pp. 67–70.
604. Paven, M., Papadopoulos, P., Mammen, L., Deng, X., Sachdev, H., Vollmer, D., and Butt, H.-J. (2014). Optimization of superamphiphobic layers based on candle soot, *Pure Appl. Chem.*, **86**, pp. 87–96.
605. Liang, C.-J., Liao, J.-D., Li, A.-J., Chen, C., Lin, H.-Y., Wang, X.-J., and Xu, Y.-H. (2014). Relationship between wettabilities and chemical compositions of candle soots, *Fuel*, **128**, pp. 422–427.
606. He, J., Li, H., Liu, X., and Qu, M. (2013). Fabricating superamphiphobic surface with fluorosilane glued carbon nanospheres films, *J. Nanosci. Nanotechnol.*, **13**, pp. 1974–1979.
607. Yu, H., Tian, X., Luo, H., and Ma, X. (2015). Hierarchically textured surfaces of versatile alloys for superamphiphobicity, *Mater. Lett.*, **138**, pp. 184–187.
608. Esmeryan, K. D., Castano, C. E., Bressler, A. H., Abolghasemibizaki, M., and Mohammadi, R. (2016). Rapid synthesis of inherently robust and stable superhydrophobic carbon soot coatings, *Appl. Surf. Sci.*, **369**, pp. 341–347.
609. Esmeryan, K. D., Radeva, E. I., and Avramov, I. D. (2016). Durable superhydrophobic carbon soot coatings for sensor applications, *J. Phys. D Appl. Phys.*, **49**, pp. 025309/1–025309/9.
610. Deng, X., Schellenberger, F., Papadopoulos, P., Vollmer, D., and Butt, H.-J. (2013). Liquid drops impacting superamphiphobic coatings, *Langmuir*, **29**, pp. 7847–7856.
611. Draper, M. C., Crick, C. R., Orlickaite, V., Turek, V. A., Parkin, I. P., and Edel, J. B. (2013). Superhydrophobic surfaces as an on-chip microfluidic toolkit for total droplet control, *Anal. Chem.*, **85**, pp. 5405–5410.
612. Freire, S. L. S., and Tanner, B. (2013). Additive-free digital microfluidics, *Langmuir*, **29**, pp. 9024–9030.
613. Li, Y., Zhu, X., Zhou, X., Ge, B., Chen, S., and Wu, W. (2014). A facile way to fabricate a superamphiphobic surface, *Appl. Phys. A*, **115**, pp. 765–770.
614. Zhu, X., Zhang, Z., Ren, G., Men, X., Ge, B., and Zhou, X. (2014). Designing transparent superamphiphobic coatings directed by carbon nanotubes, *J. Colloid Interface Sci.*, **421**, pp. 141–145.

615. Zhang, M., Zhang, T., and Cui, T. (2011). Wettability conversion from superoleophobic to superhydrophilic on titania/single-walled carbon nanotube composite coatings, *Langmuir*, **27**, pp. 9295–9301.
616. Wang, P., Chen, M., Han, H., Fan, X., Liu, Q., and Wang, J. (2016). Transparent and abrasion-resistant superhydrophobic coating with robust self-cleaning function in either air or oil, *J. Mater. Chem. A*, **4**, pp. 7869–7874.
617. Sumino, E., Saito, T., Noguchi, T., and Sewada, H. (2015). Facile creation of superoleophobic and superhydrophilic surface by using perfluoropolyether dicarboxylic acid/silica nanocomposites, *Polym. Adv. Technol.*, **26**, pp. 345–352.
618. Zhang, Y.-Y., Ge, Q., Yang, L.-L., Shi, X.-J., Li, J.-J., Yang, D.-Q., and Sacher, E. (2015). Durable superhydrophobic PTFE films through the introduction of micro- and nanostructured pores, *Appl. Surf. Sci.*, **339**, pp. 151–157.
619. Ipekci, H. H., Arkaz, H., Onses, M. S., and Hancer, M. (2016). Superhydrophobic coatings with improved mechanical robustness based on polymer brushes, *Surf. Coat. Technol.*, **299**, pp. 162–168.
620. Shen, L., Qiu, W., Wang, W., Xiao, G., and Guo, Q. (2015). Facile fabrication of superhydrophobic conductive graphite nanoplatelet/vapor-grown carbon fiber/polypropylene composite coatings, *Compos. Sci. Technol.*, **117**, pp. 39–45.
621. Yamauchi, K., Ochiai, T., and Yamauchi, G. (2015). The synergetic antibacterial performance of a Cu/WO_3-added PTFE particulate superhydrophobic composite material, *J. Biomater. Nanobiotechnol.*, **6**, pp. 1–7.
622. Chang, Y.-C., Lee, C.-C., Huang, S.-R., Kuo, C.-C., and Wei, H.-S. (2016). An easy and effective method to prepare superhydrophobic inorganic/organic thin film and improve mechanical property, *Thin Solid Films*, **618**, pp. 219–223.
623. Srinivasan, S., Chhatre, S. S., Mabry, J. M., Cohen, R. E., and McKinley, G. H. (2011). Solution spraying of poly(methyl methacrylate) blends to fabricate microtextured, superoleophobic surfaces, *Polymer*, **52**, pp. 3209–3218.
624. de Francisco, R., Hoyos, M., García, N., and Tiemblo, P. (2015). Superhydrophobic and highly luminescent polyfluorene/silica hybrid coatings deposited onto glass and cellulose-based substrates, *Langmuir*, **31**, pp. 3718–3726.

625. Yang, S., Wang, L., Wang, C.-F., Chen, L., and Chen, S. (2010). Superhydrophobic thermoplastic polyurethane films with transparent/fluorescent performance, *Langmuir*, **26**, pp. 18454–18458.

626. Kim, A., Ryu, S.-J., and Jung, H. (2016). Photoluminescent and superhydrophobic [Eu(Phen)$_2$]$^{3+}$-laponite/polypropylene film for long-term fluorescence stability under conditions of high humidity, *Adv. Mater. Interfaces*, **3**, p. 1500449.

627. Zhang, X., Guo, Y., Zhang, Z., and Zhang, P. (2013). Self-cleaning superhydrophobic surface based on titanium dioxide nanowires combined with polydimethylsiloxane, *Appl. Surf. Sci.*, **284**, pp. 319–323.

628. Park, E. J., Kim, K.-D., Yoon, H. S., Jeong, M.-G., Kim, D. H., Lim, D. C., Kim, Y. H., and Kim, Y. D. (2014). Fabrication of conductive, transparent and superhydrophobic thin films consisting of multi-walled carbon nanotubes, *RSC Adv.*, **4**, pp. 30368–30374.

629. Mokarian, Z., Rasuli, R., and Abedini, Y. (2016). Facile synthesis of stable superhydrophobic nanocomposite based on multi-walled carbon nanotubes, *Appl. Surf. Sci.*, **369**, pp. 567–575.

630. Pan, Z., Wang, T., Sun, S., and Zhao, B. (2016). Durable microstructured surfaces: Combining electrical conductivity with superoleophobicity, *ACS Appl. Mater. Interfaces*, **8**, pp. 1795–1804.

631. He, Z., Ma, M., Lan, X., Chen, F., Wang, K., Deng, H., Zhang, Q., and Fu, Q. (2011). Fabrication of a transparent superamphiphobic coating with improved stability, *Soft Matter*, **7**, pp. 6435–6443.

632. Ejenstam, L., Swerin, A., and Claesson, P. M. (2016). Toward superhydrophobic polydimethylsiloxane–silica particle coatings, *J. Dispers. Sci. Technol.*, **37**, pp. 1375–1383.

633. Basu, B. J., Kumar, V. D., and Anandan, C. (2012). Surface studies on superhydrophobic and oleophobic polydimethylsiloxane-silica nanocomposite coating system, *Appl. Surf. Sci.*, **261**, pp. 807–814.

634. Basu, B. J., Bharathidasan, T., and Anandan, C. (2013). Superhydrophobic oleophobic PDMS-silica nanocomposite coating, *Surf. Innovations*, **1**, pp. 40–51.

635. Bharathidasan, T., Narayanan, T. N., Sathyanaryanan, S., and Sreejakumari, S. S. (2015). Above 170° water contact angle and oleophobicity of fluorinated graphene oxide based transparent polymeric films, *Carbon*, **84**, pp. 207–213.

636. Neelakantan, N. K., Weisensee, P. B., Overcash, J. W., Torrealba, E. J., King, W. P., and Suslick, K. S. (2015). Spray-on omniphobic ZnO coatings, *RSC Adv.*, **5**, pp. 69243–69250.

637. Ge, B., Zhang, Z., Men, X., Zhu, X., and Zhou, X. (2014). Sprayed superamphiphobic coatings on copper substrate with enhanced corrosive resistance, *Appl. Surf. Sci.*, **293**, pp. 271–274.
638. Campos, R., Guenthner, A. J., Meuler, A. J., Tuteja, A., Cohen, R. E., McKinley, G. H., Haddad, T. S., and Mabry, J. M. (2012). Superoleophobic surfaces through control of sprayed-on stochastic topography, *Langmuir*, **28**, pp. 9834–9841.
639. Zheng, Y., Mo, C., Wang, F., and Mo, Q. (2015). Facile approach in fabricating TiO_2/organic functional coatings, *Int. J. Electrochem. Sci.*, **10**, pp. 10344–10354.
640. Wang, X., Hu, H., Ye, Q., Gao, T., Zhou, F., and Xue, Q. (2012). Superamphiphobic coatings with coralline-like structure enabled by one-step spray of polyurethane/carbon nanotube composites, *J. Mater. Chem.*, **22**, pp. 9624–9631.
641. Zhang, L., Zha, D.-A., Du, T., Mei, S., Shi, Z., and Jin, Z. (2011). Formation of superhydrophobic microspheres of poly(vinylidene fluoride-hexafluoropropylene)/graphene composite via gelation, *Langmuir*, **27**, pp. 8943–8949.
642. Hsieh, C.-T., Wu, F.-L., and Chen, W.-Y. (2009). Super water- and oil-repellencies from silica-based nanocoatings, *Surf. Coat. Technol.*, **203**, pp. 3377–3384.
643. Hsieh, C.-T., Chang, B.-S., and Lin, J.-Y. (2011). Improvement of water and oil repellency on wood substrates by using fluorinated silica nanocoating, *Appl. Surf. Sci.*, **257**, pp. 7997–8002.
644. Steele, A., Bayer, I., and Loth, E. (2009). Inherently superoleophobic nanocomposite coatings by spray atomization, *Nano Lett.*, **9**, pp. 501–505.
645. Das, A., Schutzius, T. M., Bayer, I. S., and Megaridis, C. M. (2012). Superoleophobic and conductive carbon nanofiber/fluoropolymer composite films, *Carbon*, **50**, pp. 1346–1354.
646. Schutzius, T. M., Elsharkawy, M., Tiwari, M. K., and Megaridis, C. M. (2012). Surface tension confined (STC) tracks for capillary-driven transport of low surface tension liquids, *Lab Chip*, **12**, pp. 5237–5242.
647. Wang, C.-F., Hung, S.-W., Kuo, S.-W., and Chang, C.-J. (2014). Combining hierarchical surface roughness with fluorinated surface chemistry to preserve superhydrophobicity after organic contamination, *Appl. Surf. Sci.*, **320**, pp. 658–663.
648. Lakshmi, R. V., Bharathidasan, T., Bera, P., and Basu, B. J. (2012). Fabrication of superhydrophobic and oleophobic sol-gel nanocomposite coating, *Surf. Coat. Technol.*, **206**, pp. 3888–3894.

649. Muthiah, P., Bhushan, B., Yun, K., and Kondo, H. (2013). Dual-layered-coated mechanically-durable superomniphobic surfaces with anti-smudge properties, *J. Colloid Interface Sci.*, **409**, pp. 227–236.
650. Cengiz, U., and Cansoy C. E. (2015). Applicability of Cassie–Baxter equation for superhydrophobic fluoropolymer–silica composite films, *Appl. Surf. Sci.*, **335**, pp. 99–106.
651. Xiong, D., Liu, G., Hong, L., and Duncan, E. J. S. (2011). Superamphiphobic diblock copolymer coatings, *Chem. Mater.*, **23**, pp. 4357–4366.
652. Zhang, G., Lin, S., Wyman, I., Zou, H., Hu, J., Liu, G., Wang, J., Li, F., Liu, F., and Hu, M. (2013). Omniphobic low moisture permeation transparent polyacrylate/silica nanocomposite, *ACS Appl. Mater. Interfaces*, **5**, pp. 2991–2998.
653. Zhang, G., Hu, J., Liu, G., Zou, H., Tu, Y., Li, F., Hu, S., and Luo, H. (2013). Bi-functional random copolymers for one-pot fabrication of superamphiphobic particulate coatings, *J. Mater. Chem. A*, **1**, pp. 6226–6237.
654. Xiong, D., Liu, G., and Duncan, E. J. S. (2013). Robust amphiphobic coatings from bi-functional silica particles on flat substrates, *Polymer*, **54**, pp. 3008–3016.
655. Jiang, W., Grozea, C. M., Shi, Z., and Liu, G. (2014). Fluorinated raspberry-like polymer particles for superamphiphobic coatings, *ACS Appl. Mater. Interfaces*, **6**, pp. 2629–2638.
656. Gao, D., and Jia, M. (2015). Hierarchical ZnO particles grafting by fluorocarbon polymer derivative: Preparation and superhydrophobic behavior, *Appl. Surf. Sci.*, **343**, pp. 172–180.
657. Lee, S. G., Ham, D. S., Lee, D. Y., Bong, H., and Cho, K. (2013). Transparent superhydrophobic/translucent superamphiphobic coatings based on silica–fluoropolymer hybrid nanoparticles, *Langmuir*, **29**, pp. 15051–15057.
658. Jiang, C., Wang, Q., and Wang, T. (2013). Reversible switching between hydrophobicity and oleophobicity of polyelectrolyte-functionalized multiwalled carbon nanotubes via counterion exchange, *New J. Chem.*, **37**, pp. 810–814.
659. Xiong, D., Liu, G., Zhang, J., and Duncan, S. (2011). Bifunctional core–shell–corona particles for amphiphobic coatings, *Chem. Mater.*, **23**, pp. 2810–2820.
660. Xiong, L., Kendrick, L. L., Heusser, H., Webb, J. C., Sparks, B. J., Goetz, J. T., Guo, W., Stafford, C. M., Blanton, M. D., Nazarenko, S., and Patton, D. L. (2014). Spray-deposition and photopolymerization of

organic–inorganic thiol ene resins for fabrication of superamphiphobic surfaces, *ACS Appl. Mater. Interfaces*, **6**, pp. 10763–10774.
661. Sparks, B. J., Hoff, E. F. T., Xiong, L., Goetz, J. T., and Patton, D. L. (2013). Superhydrophobic hybrid inorganic–organic thiol ene surfaces fabricated via spray-deposition and photopolymerization, *ACS Appl. Mater. Interfaces*, **5**, 1811–1817.
662. Li, F., Du, M., and Zheng, Q. (2016). Dopamine/silica nanoparticle assembled, microscale porous structure for versatile super-amphiphobic coating, *ACS Nano*, **10**, pp. 2910–2921.
663. Nishizawa, S., and Shiratori, S. (2012). Water-based preparation of highly oleophobic thin films through aggregation of nanoparticles using layer-by-layer treatment, *Appl. Surf. Sci.*, **263**, pp. 8–13.
664. Hong, J., and Kang, S. W. (2011). Hydrophobic properties of colloidal films coated with multi-wall carbon nanotubes/reduced graphene oxide multilayers, *Colloids Surf. A*, **374**, pp. 54–57.
665. Yoon, M., Kim, Y., and Cho, J. (2011). Multifunctional colloids with optical, magnetic, and superhydrophobic properties derived from nucleophilic substitution-induced layer-by-layer assembly in organic media, *ACS Nano*, **5**, pp. 5417–5426.
666. Wang, J.-L., Ren, K.-F., Chang, H., Zhang, S.-M., Jin, L.-J., and Ji, J. (2014). Facile fabrication of robust superhydrophobic multilayered film based on bioinspired poly(dopamine)-modified carbon nanotubes, *Phys. Chem. Chem. Phys.*, **16**, pp. 2936–2943.
667. Xu, H., Hu, M., Ren, K.-F., Wang, J.-L., Liu, X.-S., Jia, F., Zhao, Y.-X., and Ji, J. (2016). Spraying layer-by-layer assembly film based on the coordination bond of bioinspired polydopamine–FeIII, *Thin Solid Films*, **600**, pp. 76–82.
668. Wang, J., Zhang, Y., Wang, S., Song, Y., and Jiang, L. (2011). Bioinspired colloidal photonic crystals with controllable wettability, *Acc. Chem. Res.*, **44**, pp. 405–415.
669. Zhang, J., and Yang, B. (2010). Patterning colloidal crystals and nanostructure arrays by soft lithography, *Adv. Funct. Mater.*, **20**, pp. 3411–3424.
670. Raza, M. A., Kooij, E. S., van Silfhout, A., and Poelsema, B. (2010). Superhydrophobic surfaces by anomalous fluoroalkylsilane self-assembly on silica nanosphere arrays, *Langmuir*, **26**, pp. 12962–12972.
671. Hsieh, C.-T., Wu, F.-L., and Chen, W.-Y. (2010). Superhydrophobicity and superoleophobicity from hierarchical silica sphere stacking layers, *Mater. Chem. Phys.*, **121**, pp. 14–21.

672. Xu, Q. F., Wang, J. N., and Sanderson, K. D. (2010). A general approach for superhydrophobic coating with strong adhesion strength, *J. Mater. Chem.*, **20**, pp. 5961–5966.
673. Yang, H., Dou, X., Fang, Y., and Jiang, P. (2013). Self-assembled biomimetic superhydrophobic hierarchical arrays, *J. Colloid Interface Sci.*, **405**, pp. 51–57.
674. Tsai, H.-J., and Lee, Y.-L. (2007). Facile method to fabricate raspberry-like particulate films for superhydrophobic surfaces, *Langmuir*, **23**, pp. 12687–12692.
675. Raza, M. A., Kooij, E. S., van Silfhout, A., Zandvliet, H. J. W., and Poelsema, B. (2012). A colloidal route to fabricate hierarchical sticky and non-sticky substrates, *J. Colloid Interface Sci.*, **385**, pp. 73–80.
676. Raza, M. A., Zandvliet, H. J. W., Poelsema, B., and Kooij, E. S. (2015). Hydrophobic surfaces with tunable dynamic wetting properties via colloidal assembly of silica microspheres and gold nanoparticles, *J. Sol-Gel Sci. Technol.*, **74**, pp. 357–367.
677. Xiu, Y., Zhu, L., Hess, D. W., and Wong, C. P. (2006). Biomimetic creation of hierarchical surface structures by combining colloidal self-assembly and Au sputter deposition, *Langmuir*, **22**, pp. 9676–9681.
678. Hong, J., Bae, W. K., Lee, H., Oh, S., Char, K., Caruso, F., and Cho, J. (2007). Tunable superhydrophobic and optical properties of colloidal films coated with block-copolymer-micelles/micelle-multilayers, *Adv. Mater.*, **19**, pp. 4364–4369.
679. Horiuchi, Y., Fujiwara, K., Kamegawa, T., Mori, K., and Yamashita, H. (2011). An efficient method for the creation of a superhydrophobic surface: Ethylene polymerization over self-assembled colloidal silica nanoparticles incorporating single-site Cr-oxide catalysts, *J. Mater. Chem.*, **21**, pp. 8543–8546.
680. Min, W.-L., Jiang, P., and Jiang, B. (2008). Large-scale assembly of colloidal nanoparticles and fabrication of periodic subwavelength structures, *Nanotechnology*, **19**, pp. 475604/1–475604/7.
681. Van, T. N., Lee, Y. K., Lee, J., and Park, J. Y. (2013). Tuning hydrophobicity of TiO_2 layers with silanization and self-assembled nanopatterning, *Langmuir*, **29**, pp. 3054–3060.
682. Zhan, Y., Zhao, J., Lu, W., Yang, B., Wei, J., and Yu, Y. (2015). Biomimetic submicroarrayed cross-linked liquid crystal polymer films with different wettability via colloidal lithography, *ACS Appl. Mater. Interfaces*, **7**, pp. 25522–25528.

683. Teli, M. D., and Annaldewar, B. N. (2017). Superhydrophobic and ultraviolet protective nylon fabrics by modified nano silica coating, *J. Text. Inst.*, **108**, pp. 460–466.

684. Mahadik, S. A., Pedraza, F. D., Relekar, B. P., Parale, V. G., Lohar, G. M., and Thorat, S. S. (2016). Synthesis and characterization of superhydrophobic–superoleophilic surface, *J. Sol-Gel Sci. Technol.*, **78**, pp. 475–481.

685. Xu, Z., Zhao, Y., Wang, H., Zhou, H., Qin, C., Wang, X., and Lin, T. (2016). Fluorine-free superhydrophobic coatings with pH-induced wettability transition for controllable oil–water separation, *ACS Appl. Mater. Interfaces*, **8**, pp. 5661–5667.

686. Nateghi, M. R., and Shateri-Khalilabad, M. (2015). Silver nanowire-functionalized cotton fabric, *Carbohydr. Polym.*, **117**, pp. 160–168.

687. Wu, M., Ma, B., Pan, T., Chen, S., and Sun, J. (2016). Silver-nanoparticle-colored cotton fabrics with tunable colors and durable antibacterial and self-healing superhydrophobic properties, *Adv. Funct. Mater.*, **26**, pp. 569–576.

688. Das, I., and De, G. (2015). Zirconia based superhydrophobic coatings on cotton fabrics exhibiting excellent durability for versatile use, *Sci. Rep.*, **5**, p. 18503.

689. Gonçalves, A. G., Jarrais, B., Pereira, C., Morgado, J., Freire, C., and Pereira, M. F. R. (2012). Functionalization of textiles with multi-walled carbon nanotubes by a novel dyeing-like process, *J. Mater. Sci.*, **47**, pp. 5263–5275.

690. Shang, Y., Si, Y., Raza, A., Yang, L., Mao, X., Ding, B., and Yu, J. (2012). An in situ polymerization approach for the synthesis of superhydrophobic and superoleophilic nanofibrous membranes for oil–water separation, *Nanoscale*, **4**, pp. 7847–7854.

691. Feng, S., Zhong, Z., Zhang, F., Wang, Y., and Xing, W. (2016). Amphiphobic polytetrafluoroethylene membranes for efficient organic aerosol removal, *ACS Appl. Mater. Interfaces*, **8**, pp. 8773–8781.

692. Bayer, I. S., Fragouli, D., Attanasio, A., Sorce, B., Bertoni, G., Brescia, R., Di Corato, R., Pellegrino, T., Kalyva, M., Sabella, S., Pompa, P. P., Cingolani, R., and Athanassiou, A. (2011). Water-repellent cellulose fiber networks with multifunctional properties, *ACS Appl. Mater. Interfaces*, **3**, pp. 4024–4031.

693. Liu, W., Miao, P., Xiong, L., Du, Y., Han, X., and Xu, P. (2014). Superhydrophobic Ag nanostructures on polyaniline membranes with strong SERS enhancement, *Phys. Chem. Chem. Phys.*, **16**, pp. 22867–22873.

694. Li, B., Liu, X., Zhang, X., Zou, J., Chai, W., and Lou, Y. (2015). Rapid adsorption for oil using superhydrophobic and superoleophilic polyurethane sponge, *J. Chem. Technol. Biotechnol.*, **90**, pp. 2106–2112.
695. Yu, L., Hao, G., Zhou, S., and Jiang, W. (2016). Durable and modified foam for cleanup of oil contamination and separation of oil–water mixtures, *RSC Adv.*, **6**, pp. 24773–24779.
696. Yan, L., Li, J., Li, W., Zha, F., Feng, H., and Hu, D. (2016). A photo-induced ZnO coated mesh for on-demand oil/water separation based on switchable wettability, *Mater. Lett.*, **163**, pp. 247–249.
697. Li, J., Kang, R., Tang, X., She, Y., Yang, H., and Zha, F. (2016). Superhydrophobic meshes that can repel hot water and strong corrosive liquids used for efficient gravity-driven oil/water separation, *Nanoscale*, **8**, pp. 7638–7645.
698. Budunoglu, H., Yildirim, A., Guler, M. O., and Bayindir, M. (2011). Highly transparent, flexible, and thermally stable superhydrophobic ORMOSIL aerogel thin films, *ACS Appl. Mater. Interfaces*, **3**, pp. 539–545.
699. Yildirim, A., Khudiyev, T., Daglar, B., Budunoglu, H., Okyay, A. K., and Bayindir, M. (2013). Superhydrophobic and omnidirectional antireflective surfaces from nanostructured Ormosil colloids, *ACS Appl. Mater. Interfaces*, **5**, pp. 853–860.
700. Yu, H., Liang, X., Wang, J., Wang, M., and Yang, S. (2015). Preparation and characterization of hydrophobic silica aerogel sphere products by co-precursor method, *Solid State Sci.*, **48**, pp. 155–162.
701. Yu, Y., Wu, X., Guo, D., and Fang, J. (2014). Preparation of flexible, hydrophobic, and oleophilic silica aerogels based on a methyltriethoxysilane precursor, *J. Mater. Sci.*, **49**, pp. 7715–7722.
702. Yu, Y., Wu, X., and Fang, J. (2015). Superhydrophobic and superoleophilic "sponge-like" aerogels for oil/water separation, *J. Mater. Sci.*, **50**, pp. 5115–5124.
703. Rodriguez, J. E., Anderson, A. M., and Carroll, M. K. (2014). Hydrophobicity and drag reduction properties of surfaces coated with silica aerogels and xerogels, *J. Sol-Gel Sci. Technol.*, **71**, pp. 490–500.
704. Nadargi, D., Gurav, J., Marioni, M. A., Romer, S., Matam S., and Koebel, M. M. (2015). Methyltrimethoxysilane (MTMS)-based silica-iron oxide superhydrophobic nanocomposites, *J. Colloid Interface Sci.*, **459**, pp. 123–126.
705. Yun, S., Luo, H., and Gao, Y. (2014). Ambient-pressure drying synthesis of large resorcinol–formaldehyde-reinforced silica aerogels with enhanced mechanical strength and superhydrophobicity, *J. Mater. Chem. A*, **2**, pp. 14542–14549.

706. Li, S. H., Lin, W. G., Huang, B. C., Wang, L.-J., Gu, W.-B., Wang, W.-M., Yang, Z. Y., Wang, Y., and Zhu, J. H. (2016). Capturing nitrosamines in aqueous solution by composited super-hydrophobic silicic xerogel, *Microporous Mesoporous Mater.*, **227**, pp. 161–168.

707. Goswami, D., Medda, S. K., and De, G. (2011). Superhydrophobic films on glass surface derived from trimethylsilanized silica gel nanoparticles, *ACS Appl. Mater. Interfaces*, **3**, pp. 3440–3447.

708. Manca, M., Cannavale, A., De Marco, L., Aricò, A. S., Cingolani, R., and Gigli, G. (2009). Durable superhydrophobic and antireflective surfaces by trimethylsilanized silica nanoparticles-based sol–gel processing, *Langmuir*, **25**, pp. 6357–6362.

709. Amirkhani, L., Moghaddas, J., and Jafarizadeh-Malmiri, H. (2016). Candida rugosa lipase immobilization on magnetic silica aerogel nanodispersion, *RSC Adv.*, **6**, pp. 12676–12687.

710. He, J., Li, X., Su, D., Ji, H., Zhang, X., and Zhang, W. (2016). Super-hydrophobic hexamethyl-disilazane modified ZrO_2–SiO_2 aerogels with excellent thermal stability, *J. Mater. Chem. A*, **4**, pp. 5632–5638.

711. Lin, J., Chen, H., Fei, T., Liu, C., and Zhang, J. (2013). Highly transparent and thermally stable superhydrophobic coatings from the deposition of silica aerogels, *Appl. Surf. Sci.*, **273**, pp. 776–786.

712. Jin, H., Tian, X., Ikkala, O., and Ras, R. H. A. (2013). Preservation of superhydrophobic and superoleophobic properties upon wear damage, *ACS Appl. Mater. Interfaces*, **5**, pp. 485–488.

713. Zou, F., Peng, L., Fu, W., Zhang, J., and Li, Z. (2015). Flexible superhydrophobic polysiloxane aerogels for oil–water separation via one-pot synthesis in supercritical CO_2, *RSC Adv.*, **5**, pp. 76346–76351.

714. Wang, X., and Jana, S. C. (2013). Synergetic hybrid organic–inorganic aerogels, *ACS Appl. Mater. Interfaces*, **5**, pp. 6423–6429.

715. Matias, T., Varino, C., de Sousa, H. C., Braga, M. E. M., Portugal, A., Coelho, J. F. J., and Duraes, L. (2016). Novel flexible, hybrid aerogels with vinyl- and methyltrimethoxysilane in the underlying silica structure, *J. Mater. Sci.*, **51**, pp. 6781–6792.

716. Guo, P., Zhai, S.-R., Xiao, Z.-Y., Zhang, F., An, Q.-D., and Song, X.-W. (2014). Preparation of superhydrophobic materials for oil/water separation and oil absorption using PMHS–TEOS-derived xerogel and polystyrene, *J. Sol-Gel Sci. Technol.*, **72**, pp. 385–393.

717. Sanli, D., and Erkey, C. (2013). Monolithic composites of silica aerogels by reactive supercritical deposition of hydroxy-terminated poly(dimethylsiloxane), *ACS Appl. Mater. Interfaces*, **5**, pp. 11708–11717.

718. Kim, H. M., Noh, Y. J., Yu, J., Kim, S. Y., and Youn, J. R. (2015). Silica aerogel/polyvinyl alcohol (PVA) insulation composites with preserved aerogel pores using interfaces between the superhydrophobic aerogel and hydrophilic PVA solution, *Compos. Part A*, **75**, pp. 39–45.

719. Lin, Y., Ehlert, G. J., Bukowsky, C., and Sodano, H. A. (2011). Superhydrophobic functionalized graphene aerogels, *ACS Appl. Mater. Interfaces*, **3**, pp. 2200–2203.

720. Ren, H., Shi, X., Zhu, J., Zhang, Y., Bi, Y., and Zhang, L. (2016). Facile synthesis of N-doped graphene aerogel and its application for organic solvent adsorption, *J. Mater. Sci.*, **51**, pp. 6419–6427.

721. Li, R., Chen, C., Li, J., Xu, L., Xiao, G., and Yan D. (2014). A facile approach to superhydrophobic and superoleophilic graphene/polymer aerogels, *J. Mater. Chem. A*, **2**, pp. 3057–3064.

722. Ha, H., Shanmuganathan, K., and Ellison, C. J. (2015). Mechanically stable thermally crosslinked poly(acrylic acid)/reduced graphene oxide aerogels, *ACS Appl. Mater. Interfaces*, **7**, pp. 6220–6229.

723. Hu, H., Zhao, Z., Wan, W., Gogotsi, Y., and Qiu, J. (2014). Polymer/graphene hybrid aerogel with high compressibility, conductivity, and "sticky" superhydrophobicity, *ACS Appl. Mater. Interfaces*, **6**, pp. 3242–3249.

724. Liu, W., Cai, J., Ding, Z., and Li, Z. (2015). TiO_2/RGO composite aerogels with controllable and continuously tunable surface wettability for varied aqueous photocatalysis, *Appl. Catalys. B*, **174–175**, pp. 421–426.

725. Hu, H., Zhao, Z., Gogotsi, Y., and Qiu, J. (2014). Compressible carbon nanotube–graphene hybrid aerogels with superhydrophobicity and superoleophilicity for oil sorption, *Environ. Sci. Technol. Lett.*, **1**, pp. 214–220.

726. Li, L., Li, B., and Zhang, J. (2016). Dopamine-mediated fabrication of ultralight graphene aerogels with low volume shrinkage, *J. Mater. Chem. A*, **4**, pp. 512–518.

727. Zhang, J., Li, B., Li, L., and Wang, A. (2016). Ultralight, compressible and multifunctional carbon aerogels based on natural tubular cellulose, *J. Mater. Chem. A*, **4**, pp. 2069–2074.

728. Li, Z., Tang, X.-Z., Zhu, W., Thompson, B. C., Huang, M., Yang, J., Hu, X., and Khor, K. A. (2016). Single-step process towards achieving superhydrophobic reduced graphene oxide, *ACS Appl. Mater. Interfaces*, **8**, pp. 10985–10994.

729. Zhou, S., Zhou, X., Jiang, W., Wang, T., Zhang, N., Lu, Y., Yu, L., and Yin, Z. (2016). (3-Mercaptopropyl)trimethoxysilane-assisted

synthesis of macro- and mesoporous graphene aerogels exhibiting robust superhydrophobicity and exceptional thermal stability, *Ind. Eng. Chem. Res.*, **55**, pp. 948–953.

730. Li, Y., Zhu, X., Ge, B., Men, X., Li, P., and Zhang, Z. (2015). Versatile fabrication of magnetic carbon fiber aerogel applied for bidirectional oil–water separation, *Appl. Phys. A*, **120**, pp. 949–957.

731. Pham, T., Goldstein, A. P., Lewicki, J. P., Kucheyev, S. O., Wang, C., Russell, T. P., Worsley, M. A., Woo, L., Mickelson, W., and Zettl, A. (2015). Nanoscale structure and superhydrophobicity of sp^2-bonded boron nitride aerogels, *Nanoscale*, **7**, pp. 10449–10458.

732. Qu, L., Dai, L., Stone, M., Xia, Z., and Wang, Z. L. (2008). Carbon nanotube arrays with strong shear binding-on and easy normal lifting-off, *Science*, **322**, pp. 238–242.

733. Ge, L., Sethi, S., Ci, L., Ajayan, P. M., and Dhinojwala, A. (2007). Carbon nanotube-based synthetic gecko tapes, *Proc. Natl. Acad. Sci. U. S. A.*, **104**, pp. 10792–10795.

734. Ozden, S., Ge, L., Narayanan, T. N., Hart, A. H. C., Yang, H., Sridhar, S., Vajtai, R., and Ajayan, P. M. (2014). Anisotropically functionalized carbon nanotube array based hygroscopic scaffolds, *ACS Appl. Mater. Interfaces*, **6**, pp. 10608–10613.

735. Qu, L., and Dai, L. (2007). Gecko-foot-mimetic aligned single-walled carbon nanotube dry adhesives with unique electrical and thermal properties, *Adv. Mater.*, **19**, pp. 3844–3849.

736. Nair, H., Tiggelaar, R. M., Thakur, D. B., Gardeniers, J. G. E., van Houselt, A., and Lefferts, L. (2013). Evidence of wettability variation on carbon nanofiber layers grown on oxidized silicon substrates, *Chem. Eng. J.*, **227**, pp. 56–65.

737. Man, Y., Chen, Z., Zhang, Y., and Guo, P. (2016). Synthesis and characterization of novel carbon nanotube array supported Fe$_3$C nanocomposites with honeycomb structures, *Mater. Des.*, **97**, pp. 417–423.

738. De Nicola, F., Castrucci, P., Scarselli, M., Nanni, F., Cacciotti, I., and De Crescenzi, M. (2015). Super-hydrophobic multi-walled carbon nanotube coatings for stainless steel, *Nanotechnology*, **26**, pp. 145701/1–145701/6.

739. Ming, Z., Jian, L., Chunxia, W., Xiaokang, Z., and Lan, C. (2011). Fluid drag reduction on superhydrophobic surfaces coated with carbon nanotube forests (CNTs), *Soft Matter*, **7**, pp. 4391–4396.

740. De Nicola, F., Castrucci, P., Scarselli, M., Nanni, F., Cacciotti, I., and De Crescenzi, M. (2013). Dry oxidation and vacuum annealing

treatments for tuning the wetting properties of carbon nanotube arrays, *J. Vis. Exp.*, **74**, p. e50378.

741. Ramos, S. C., Vasconcelos, G., Antunes, E. F., Lobo, A. O., Trava-Airoldi, V. J., and Corat, E. J. (2010). Total re-establishment of superhydrophobicity of vertically-aligned carbon nanotubes by CO_2 laser treatment, *Surf. Coat. Technol.*, **204**, pp. 3073–3077.

742. Ramos, S. C., Vasconcelos, G., Antunes, E. F., Lobo, A. O., Trava-Airoldi, V. J., and Corat, E. J. (2010). Wettability control on vertically-aligned multi-walled carbon nanotube surfaces with pulsed DC plasma and CO_2 treatments, *Diam. Rel. Mater.*, pp. 752–755.

743. Panagiotopoulos, N. T., Diamanti, E. K., Koutsokeras, L. E., Baikousi, M., Kordatos, E., Matikas, T. E., Gournis, D., and Patsalas, P. (2012). Nanocomposite catalysts producing durable, super-black carbon nanotube systems: Applications in solar thermal harvesting, *ACS Nano*, **6**, pp. 10475–10485.

744. Zhang, Y., Stan, L., Xu, P., Wang, H.-L., Doorn, S. K., Htoon, H., Zhu, Y., and Jia, Q. (2009). A double-layered carbon nanotube array with super-hydrophobicity, *Carbon*, **47**, pp. 3332–3336.

745. Li, H., Wang, X., Song, Y., Liu, Y., Li, Q., Jiang, L., and Zhu, D. (2001). Super-"amphiphobic" aligned carbon nanotube films, *Angew. Chem. Int. Ed.*, **40**, pp. 1743–1746.

746. Lau, K. K. S., Bico, J., Teo, K. B. K., Chhowalla, M., Amaratunga, G. A. J., Milne, W. I., McKinley, G. H., and Gleason, K. K. (2003). Superhydrophobic carbon nanotube forests, *Nano Lett.*, **3**, pp. 1701–1705.

747. Babu, D. J., Varanakkottu, S. N., Eifert, A., de Koning, D., Cherkashinin, G., Hardt, S., and Schneider, J. J. (2014). Inscribing wettability gradients onto superhydrophobic carbon nanotube surfaces, *Adv. Mater. Interfaces*, **1**, p. 1300049.

748. Huang, L., Lau, S. P., Yang, H. Y., Leong, E. S. P., and Yu, S. F. (2005). Stable superhydrophobic surface via carbon nanotubes coated with a ZnO thin film, *J. Phys. Chem. B*, **109**, pp. 7746–7748.

749. Tang, Y., Gou, J., and Hu, Y. (2013). Covalent functionalization of carbon nanotubes with polyhedral oligomeric silsequioxane for superhydrophobicity and flame retardancy, *Polym. Eng. Sci.*, **53**, pp. 1021–1030.

750. Silva, T. A., Zanin, H., Saito, E., Medeiros, R. A., Vicentini, F. C., Corat, E. J., and Fatibello-Filho O. (2014). Electrochemical behaviour of vertically aligned carbon nanotubes and graphene oxide nanocomposite as electrode material, *Electrochim. Acta*, **119**, pp. 114–119.

751. Silva, T. A., Zanin, H., Vicentini, F. C., Corat, E. J., and Fatibello-Filho, O. (2014). Differential pulse adsorptive stripping voltammetric determination of nanomolar levels of atorvastatin calcium in pharmaceutical and biological samples using a vertically aligned carbon nanotube/graphene oxide electrode, *Analyst*, **139**, pp. 2832–2841.

752. Jeong, D.-W., Shin, U.-H., Kim, J. H., Kim, S.-H., Lee, H. W., and Kim, J.-M. (2014). Stable hierarchical superhydrophobic surfaces based on vertically aligned carbon nanotube forests modified with conformal silicone coating, *Carbon*, **79**, pp. 442–449.

753. Sun, T., Wang, G., Liu, H., Feng, L., Jiang, L., and Zhu, D. (2003). Control over the wettability of an aligned carbon nanotube film, *J. Am. Chem. Soc.*, **125**, pp. 14996–14997.

754. Chen, L., Xiao, Z., Chan, P. C. H., and Lee, Y.-K. (2010). Static and dynamic characterization of robust superhydrophobic surfaces built from nano-flowers on silicon micro-post arrays, *J. Micromech. Microeng.*, **20**, pp. 105001/1–105001/8.

755. Shawat, E., Perelshtein, I., Westover, A., Pint, C. L., and Nessim, G. D. (2014). Ultra high-yield one-step synthesis of conductive and superhydrophobic three-dimensional mats of carbon nanofibers via full catalysis of unconstrained thin films, *J. Mater. Chem. A*, **2**, pp. 15118–15123.

756. Meng, L.-Y., Rhee, K. Y., and Park, S.-J. (2014). Enhancement of superhydrophobicity and conductivity of carbon nanofibers-coated glass fabrics, *J. Ind. Eng. Chem.*, **20**, pp. 1672–1676.

757. Qiu, R., Zhang, Q., Wang, P., Jiang, L., Hou, J., Guo, W., and Zhang, H. (2014). Fabrication of slippery liquid-infused porous surface based on carbon fiber with enhanced corrosion inhibition property, *Colloids Surf. A*, **453**, pp. 132–141.

758. Yoon, D., Lee, C., Yun, J., Jeon, W., Cha, B. J., and Baik, S. (2012). Enhanced condensation, agglomeration, and rejection of water vapor by superhydrophobic aligned multiwalled carbon nanotube membranes, *ACS Nano*, **6**, pp. 5980–5987.

759. Jeon, W., Yun, J., Khan, F. A., and Baik, S. (2015). Enhanced water vapor separation by temperature-controlled aligned-multiwalled carbon nanotube membranes, *Nanoscale*, **7**, pp. 14316–14323.

760. Yun, J., Jeon, W., Khan, F. A., Lee, J., and Baik, S. (2015). Reverse capillary flow of condensed water through aligned multiwalled carbon nanotubes, *Nanotechnology*, **26**, pp. 235701/1–235701/10.

761. Badge, I., Sethi, S., and Dhinojwala, A. (2011). Carbon nanotube-based robust steamphobic surfaces, *Langmuir*, **27**, pp. 14726–14731.
762. Dumée, L. F., Gray, S., Duke, M., Sears, K., Schütz, J., and Finn, N. (2013). The role of membrane surface energy on direct contact membrane distillation performance, *Desalination*, **323**, pp. 22–30.
763. Román-Manso, B., Vega-Díaz, S. M., Morelos-Gómez, A., Terrones, M., Miranzo, P., and Belmonte, M. (2014). Aligned carbon nanotube/silicon carbide hybrid materials with high electrical conductivity, superhydrophobicity and superoleophilicity, *Carbon*, **80**, pp. 120–126.
764. Kousalya, A. S., Singh, K. P., and Fisher, T. S. (2015). Heterogeneous wetting surfaces with graphitic petal-decorated carbon nanotubes for enhanced flow boiling, *Int. J. Heat Mass Transfer*, **87**, pp. 380–389.
765. Vasconcellos de Siqueira Brandão, L. E., Fassini Michels, A., Camargo, K. C., Balzaretti, N. M., and Horowitz, F. (2013). Wet ability of PTFE coated diamond films, *Surf. Coat. Technol.*, **232**, pp. 384–388.
766. Cortese, B., Caschera, D., Federici, F., Ingo, G. M., and Gigli, G. (2014). Superhydrophobic fabrics for oil–water separation through a diamond like carbon (DLC) coating, *J. Mater. Chem. A*, **2**, pp. 6781–6789.
767. Stancu, E. C., Ionita, M. D., Vizireanu, S., Stanciuc, A. M., Moldovan, L., and Dinescu, G. (2010). Wettability properties of carbon nanowalls layers deposited by a radiofrequency plasma beam discharge, *Mater. Sci. Eng. B*, **169**, pp. 119–122.
768. Caschera, D., Mezzi, A., Cerri, L., de Caro, T., Riccucci, C., Ingo, G. M., Padeletti, G., Biasiucci, M., and Gigli, G. (2014). Effects of plasma treatments for improving extreme wettability behavior of cotton fabrics, *Cellulose*, **21**, pp. 741–756.
769. Dong, J., Yao, Z., Yang, T., Jiang, L., and Shen, C. (2013). Control of superhydrophilic and superhydrophobic graphene interface, *Sci. Rep.*, **3**, p. 1733.
770. Singh, E., Chen, Z., Houshmand, F., Ren, W., Peles, Y., Cheng, H.-M., and Koratkar, N. (2013). Superhydrophobic graphene foams, *Small*, **9**, pp. 75–80.
771. Verplanck, N., Galopin, E., Camart, J.-C., Thomy, V., Coffinier, Y., and Boukherroub, R. (2007). Reversible electrowetting on superhydrophobic silicon nanowires, *Nano Lett.*, **7**, pp. 813–817.
772. Lapierre, F., Thomy, V., Coffinier, Y., Blossey, R., and Boukherroub, R. (2009). Reversible electrowetting on superhydrophobic double-nanotextured surfaces, *Langmuir*, **25**, pp. 6551–6558.

773. Nguyen, T. P. N., Boukherroub, R., Thomy, V., and Coffinier, Y. (2014). Micro-and nanostructured silicon-based superomniphobic surfaces, *J. Colloid Interface Sci.*, **416**, pp. 280–288.

774. Khan, A., Sohail, S., and Jacob, C. (2015). The fabrication of stable superhydrophobic surfaces using a thin Au/Pd coating over a hydrophilic 3C-SiC nanorod network, *Appl. Surf. Sci.*, **353**, pp. 964–972.

775. Kamal, S. A. A., Ritikos, R., and Rahman, S. A. (2015). Wetting behaviour of carbon nitride nanostructures grown by plasma enhanced chemical vapour deposition technique, *Appl. Surf. Sci.*, **328**, pp. 146–153.

776. Pakdel, A., Zhi, C., Bando, Y., Nakayama, T., and Golberg, D. (2011). Boron nitride nanosheet coatings with controllable water repellency, *ACS Nano*, **5**, pp. 6507–6515.

777. Pakdel, A., Wang, X., Bando, Y., and Golberg, D. (2013). Nonwetting "white graphene" films, *Acta Mater.*, **61**, pp. 1266–1273.

778. Pakdel, A., Zhi, C., Watanabe, K., Sekiguchi, T., Nakayama, T., and Golberg, D. (2013). Nonwetting and optical properties of hexagonal boron nitride films, *Surf. Innovations*, **1**, pp. 32–39.

779. Pakdel, A., Wang, X., Zhi, C., Bando, Y., Watanabe, K., Sekiguchi, T., Nakayama, T., and Golberg, D. (2012). Facile synthesis of vertically aligned hexagonal boron nitride nanosheets hybridized with graphitic domains, *J. Mater. Chem.*, **22**, pp. 4818–4824.

780. Pakdel, A., Bando, Y., and Golberg, D. (2014). Plasma-assisted interface engineering of boron nitride nanostructure films, *ACS Nano*, **8**, pp. 10631–10639.

781. Yu, J., Qin, L., Hao, Y., Kuang, S., Bai, X., Chong, Y.-M., Zhang, W., and Wang, E. (2010). Vertically aligned boron nitride nanosheets: Chemical vapor synthesis, ultraviolet light emission, and superhydrophobicity, *ACS Nano*, **4**, pp. 414–422.

782. Cui, H., Sun, Y., and Wang, C. X. (2013). Unusual growth direction and controllable wettability of W-catalyzed AlN nanowires, *CrystEngComm*, **15**, pp. 5376–5381.

783. Li, J., Han, Q., Zhang, Y., Zhang, W., Dong, M., Besenbacher, F., Yang, R., and Wang, C. (2013). Optical regulation of protein adsorption and cell adhesion by photoresponsive GaN nanowires, *ACS Appl. Mater. Interfaces*, **5**, pp. 9816–9822.

784. Li, Y., Zheng, M., Ma, L., Zhong, M., and Shen, W. (2008). Fabrication of hierarchical ZnO architectures and their superhydrophobic surfaces with strong adhesive force, *Inorg. Chem.*, **47**, pp. 3140–3143.

785. Jayram, N. D., Sonia, S., Poongodi, S., Suresh Kumar, P., Masuda, Y., Mangalaraj, D., Ponpandian, N., and Viswanathan, C. (2015). Superhydrophobic Ag decorated ZnO nanostructured thin film as effective surface enhanced Raman scattering substrates, *Appl. Surf. Sci.*, **355**, pp. 969–977.
786. Macias-Montero, M., Borras, A., Saghi, Z., Romero-Gomez, P., Sanchez-Valencia, J. R., Gonzalez, J. C., Barranco, A., Midgley, P., Cotrino, J., and Gonzalez-Elipe, A. R. (2012). Superhydrophobic supported Ag-NPs@ZnO-nanorods with photoactivity in the visible range, *J. Mater. Chem.*, **22**, pp. 1341–1346.
787. Cao, Y., Deng, S., Hu, Q., Zhong, Q., Luo, Q.-P., Yuan, L., and Zhou, J. (2015). Three-dimensional ZnO porous films for self-cleaning ultraviolet photodetectors, *RSC Adv.*, **5**, pp. 85969–85973.
788. Yao, L., Zheng, M., Hea, S., Ma, L., Li, M., and Shen, W. (2011). Preparation and properties of ZnS superhydrophobic surface with hierarchical structure, *Appl. Surf. Sci.*, **257**, pp. 2955–2959.
789. Yadav, K., Mehta, B. R., Lakshmi, K. V., Bhattacharya, S., and Singh, J. P. (2015). Tuning the wettability of indium oxide nanowires from superhydrophobic to nearly superhydrophilic: Effect of oxygen-related defects, *J. Phys. Chem. C*, **119**, pp. 16026–16032.
790. Senthil, K., Kwak, G., and Yong, K. (2012). Fabrication of superhydrophobic vanadium pentoxide nanowires surface by chemical modification, *Appl. Surf. Sci.*, **258**, pp. 7455–7459.
791. Kwak, G., Lee, M., and Yong, K. (2010). Chemically modified superhydrophobic WOx nanowire arrays and UV photopatterning, *Langmuir*, **26**, pp. 9964–9967.
792. Lee, S., Lee, J., Park, J., Choi, Y., and Yong, K. (2012). Resistive switching WOx-Au core-shell nanowires with unexpected nonwetting stability even when submerged under water, *Adv. Mater.*, **24**, pp. 2418–2423.
793. Yadav, K., Mehta, B. R., and Singh, J. P. (2015). Superhydrophobicity and enhanced UV stability in vertically standing indium oxide nanorods, *Appl. Surf. Sci.*, **346**, pp. 361–365.
794. Zhong, M., Zheng, M., Zeng, A., and Ma, L. (2008). Direct integration of vertical In_2O_3 nanowire arrays, nanosheet chains, and photoinduced reversible switching of wettability, *Appl. Phys. Lett.*, **92**, pp. 093118/1–093118/3.
795. Gao, L. Y., Zheng, M. J., Zhong, M., Li, M., and Ma, L. (2007). Preparation and photoinduced wettability conversion of superhydrophobic β-Ga_2O_3 nanowire film, *Appl. Phys. Lett.*, **91**, pp. 013101/1–013101/3.

796. Zhou, J., Fan, J.-B., Nie, Q., and Wang, S. (2016). Three-dimensional superhydrophobic copper 7,7,8,8-tetracyanoquinodimethane biointerfaces with the capability of high adhesion of osteoblasts, *Nanoscale*, **8**, pp. 3264–3267.

797. Pan, J., Song, X., Zhang, J., Shen, H., and Xiong, Q. (2011). Switchable wettability in SnO_2 nanowires and SnO_2@SnO_2 heterostructures, *J. Phys. Chem. C*, **115**, pp. 22225–22231.

798. Hnilica, J., Schäfer, J., Foest, R., Zajíčková, L., and Kudrle, V. (2013). PECVD of nanostructured SiO_2 in a modulated microwave plasma jet at atmospheric pressure, *J. Phys. D Appl. Phys.*, **46**, pp. 335202/1–335202/8.

799. Charlton, J. J., Lavrik, N., Bradshaw, J. A., and Sepaniake, M. J. (2014). Wicking nanopillar arrays with dual roughness for selective transport and fluorescence measurements, *ACS Appl. Mater. Interfaces*, **6**, pp. 17894–17901.

Index

AAO, *see* anodized aluminum oxide
AAO membrane 42–44
absorption 76
acidic media 13, 15, 17, 19, 52, 109
acids
  carboxylic 14–16, 95, 102
  citric 45
  myristic 61
  oxalic 14, 38–39, 41, 48, 93
  stearic 15–17, 53, 93
  tetradecanoic 15
adherence 84, 86
adhesion
  high 4, 20, 97, 105–106
  low 10, 13–14, 20, 42, 71–72, 88, 97, 99, 101, 103
aerogels 93–96
  graphene oxide 95–96
  silica 93
  superhydrophobic 93–94, 96
  superhydrophobic graphene 94
  superhydrophobic silica 94
$AgNO_3$ 49–50, 63–65
ALD, *see* atomic layer deposition
alkylamine 78
AlN 99, 104
aluminum 18, 23, 27, 38, 75, 81
aluminum nanolayer 34–35
aluminum substrates 13–14, 16, 18, 27, 29, 32, 37–40, 65, 67
  anodizing 39
  bare 80
  microstructured 14
ammoniac 75–76

ammonium citrate 64–65
anodization 38–41, 43, 45–49, 109
  two-step 39, 41
anodization time 40, 47–49
anodized aluminum oxide (AAO) 26, 41, 43, 105
anodized aluminum oxide membrane 43–44
anti-icing properties 13, 40–41
anticorrosion properties 14, 16, 32, 40, 47, 53, 56–57, 62, 67, 72, 80–81
antifogging 2–3
Antipaluria urichi 5
argon 95–96, 105
atom transfer radical polymerization (ATRP) 66, 89
atomic layer deposition (ALD) 27, 70–71
atorvastatin calcium 99–100
ATRP, *see* atom transfer radical polymerization
AuZn 66
$AuZn_3$ 66

$BiCl_3$ 67
biological samples 99–100
BN, *see* boron nitride
BN nanosheets 104
BOE, *see* buffered oxide etchant

boron nitride (BN) 22, 75, 96, 104
buffered oxide etchant (BOE) 22

carbon 85, 97, 99, 101–103
carbon-based materials 101
carbon nanofibers 21, 87–88, 100
  growth of 100
carbon nanoparticles 85
carbon nanotubes 41, 79, 85, 89, 97–98
  growth of 41, 98
  multiwalled 85–86, 100–101
Cassie–Baxter equation 10–12
casting 43–44
catalyst 19, 62, 97, 99–100, 102, 104–105
chemical vapor deposition (CVD) 41, 67, 97, 99, 101, 103–106, 109
  microwave plasma 102
cicadas 5
classical superhydrophobic properties 4
CO oxidation 79–80
Co$_3$Ni 67–68
coatings 18, 31, 72, 86, 93, 100
  conductive 86–87
  fluorescent 86
  rare earth superhydrophobic 61
  superhydrophobic TiO$_2$-Cu$_2$O 24
cobalt 59, 65, 75, 79, 84
CoCl$_2$ 57, 59, 67–68
colloidal lithography 73, 90
copper 14–15, 23, 27, 29, 32–37, 47, 52, 54, 57, 66, 75, 81, 84, 100, 102
copper foams 33
copper nanocone arrays 54
crystal modifier 54, 57–59, 68

crystalline structures 32, 36, 69
crystallographic plane 71
Cu$_2$O 15, 52–53
Cu–Ni composites 55
CuO 33, 36, 66, 74–75, 79
CuO nanoflowers 33–34
CuO nanowires 36–37
CuSO$_4$ 52, 54–55, 66–67
cuticle 1, 7–8
CVD see chemical vapor deposition

deep reactive ion etching (DRIE) 22
dendritic structures 49, 60, 64–65, 67
dopamine 34–35
DRIE, see deep reactive ion etching
dynamic hydrogen bubble templating process 55–56, 60

electrochemical machining 17
electrodeposition 31, 42, 49–55, 57–62, 73, 100, 109
  metal 55
  palladium 51
  sol-gel 62
  two-step 54, 57
electrodes, nanoparticle-modified 59
electroless deposition 19, 62–67, 109
  halide-assisted 64
etching 13–19, 29, 63, 66–67, 92, 109
  chemical 15, 28, 39, 41–42, 83

electrochemical 19–20
  in acidic media 13, 15, 17, 19
  in basic media 32–33, 35, 37
ethylene glycol 57, 84, 87
ethylenediamine 64, 75–76

fabrics 92, 100
Fe 42, 58, 83, 96–97
FeCl$_3$ 15, 58–59
femtosecond lasers 27–29, 31
fluorescence properties 75
fluorinated polyacrylates 87
fluorinated SiO$_2$ nanoparticles 82, 87, 92
fluorination 34, 48, 73, 77, 79
fluorocarbon wax crystals 37

gold dendritic structures 66
graphene nanosheets 102
graphene oxide aerogel formation 96
graphite 105

HAuCl$_4$ 51, 66
HCl 13–17, 40–42, 48, 67, 93
hexadecane 14, 16, 37–41, 46–47, 64, 82–83, 85–89, 102
hexagonal-close-packed ommatidia 2
hexamethylenetetramine 69–70, 85
HNO$_3$ 14, 17, 19, 29, 37, 48, 57, 61
hydrophobic materials 1, 18–19

hydrophobic thiol 49–52, 54–55
hydrophobic waxes 3–4
hydrothermal processes 41, 68–73, 75, 77, 79–81, 109

insects 2–3, 5
iridescence properties 28

KMnO$_4$ 14

laser 27–29, 31
laser fluence 27–28, 30–31
laser parameters 29–30
laser treatment 27, 29–32
layer-by-layer (LbL) 89
layered double hydroxides (LDH) 42
LbL, *see* layer-by-layer
LDH, *see* layered double hydroxides
LEDs, *see* light-emitting diodes
light-emitting diodes (LEDs) 104
lotus leaf 2

(3-mercaptopropyl)trimethoxysilane (MPS) 96
metal ions 36, 42
metal substrate 13, 15, 17, 38, 45, 62
  non-noble 13, 32, 38
metals
  noble 49
  non-noble 52

methyltriethoxysilane (MTES) 93–94
methyltrimethoxysilane (MTMS) 93–94
microcones 22, 31, 49
microgrooves 6, 18–20, 27, 102
micropillar arrays 31, 100–101, 103
microscales 6
microspheres, nanostructured 58, 81
microstructures 13–14, 28–30, 40, 46–47, 63–64, 67
MPS, see (3-mercaptopropyl)trimethoxysilane
MTES, see methyltriethoxysilane
MTMS, see methyltrimethoxysilane
multiwalled carbon nanotubes (MWCNTs) 86, 98–101
  vertically aligned 97–99, 101
MWCNTs, see multiwalled carbon nanotubes

Na_2WO_4 60, 77
NaCl 18, 57, 75, 86
NaH_2PO_2 64–65
nanocomposites 86–87, 92
nanocones 22, 54, 57, 59
nanoflowers 33, 48, 50
nanofolds 4
nanograss 22, 33–34
nanopillars 39–40
nanoplates 64–65, 78
nanopore arrays, hexagonally packed 26, 39–40
nanopores 39–40, 43–44, 46
nanoribbons 36, 75–76
nanorod arrays 24, 28
nanorods 24–26, 28, 33, 60, 69, 71, 77–78, 103, 105–106

aligned 69–71, 106
interconnected 70
nanoroughness 13–14, 29, 83, 91
nanosecond lasers 27, 29
nanosheets 15–16, 36, 41, 50, 57, 75, 77–78, 85
nanoshells 22, 59
nanostructured superhydrophobic properties 13
nanostructures
  gold 23, 51
  helicoidal 25
  ZnO 48, 72, 78
nanowire arrays 34–35
nanowires 19–20, 33–35, 37, 48, 69, 71, 76, 78, 84–86, 102, 105–106
  aligned nickel 42
  long 74–75, 102
NaOH 33–34, 36–37, 47–49, 66–67, 75–77
NH_3 36, 64, 69, 75, 104
NH_4Cl 57, 75
NH_4F 19, 46–47
NH_4OH 36, 38, 93
$NiCl_2$ 56–58, 67–68

oil droplets 11–12
oil/water mixtures 23, 93
oil/water separation 17, 33, 37, 41, 81, 96, 101–102
organic solvents 63, 96
oxides
  metal 23, 49, 52
  nanostructured 75, 105
oxygen 32, 105

parahydrophobic property 5, 20, 105–106

PDMS, *see* polydimethylsiloxane
perfluorinated phosphate 39–40
photocatalytic activity 77, 83
photocatalytic properties 24, 46, 77–78, 96
photoluminescence 19, 48, 76–77, 85, 109
platinum 51, 66
polydimethylsiloxane (PDMS) 10, 57, 71–72, 86, 94
polydopamine 84, 89
polymer materials 43–44
polyol process 84
polystyrene spheres, colloidal 50–51, 59
polytetrafluoroethylene (PTFE) 10, 24, 61, 71, 86, 99
proteins 8
PtCl$_4$ 66
PTFE, *see* polytetrafluoroethylene

reactive ion etching (RIE) 21–22, 92
RIE, *see* reactive ion etching
rolling-spheronization granulation 48
roughness 5–6, 9, 83
  multiscale 16, 82–83

SEFS, *see* surface-enhanced fluorescence spectroscopy
self-cleaning properties 1, 14, 74–75
SERS, *see* surface-enhanced Raman scattering
SF$_6$ 21–22
SF$_6$ gas 31–32
silicon 18–19, 27, 29, 31, 65–66, 102

silicon cylindrical nanoshell arrays 22–23
silicon nanograss 21–22
silicon nanowire arrays 19–21
silicon nanowires 102
silicon wafers 18–19, 22
silicone 99–100
silver 19, 49, 63–65, 83
silver nanoparticles 19, 93, 105
  electrodeposited 50
SiO$_2$ nanoparticles 62, 82–83, 87–89, 93
SiO$_2$ nanospheres 91–92
SiO$_2$ spheres 90–91
SnCl$_2$ 60, 67
sodium dodecylsulfate 67–68
sol-gel process 62, 82, 93
solar cells 78–79
spacer lithography 22–23
sputter deposition 23–26
substrate adherence 87
substrates
  AAO 41
  alumina 29
  anodized aluminum 40
  brass 15
  copper 15, 29, 33, 36, 47, 52–53, 63, 65–68, 76, 81, 106
  copper mesh 23, 33, 48, 52, 55
  etched 13
  GaAs 64
  glass 44
  gold dendritic 66
  graphene oxide 27
  laser-treated 28
  micropattern PET 42
  micropatterned 28–30, 100
  micropillar 12
  multiwalled carbon nanotube 98
  nanoporous superhydrophobic 27
  nanostructured copper 55

nickel 56
periodic micropatterned
　copper 28
polymer 21, 34, 42
silicon 18, 21, 27, 29, 32, 65
stainless 16, 18, 30–31
steel 67
superhydrophobic 32
superhydrophobic aluminum 18
superoleophobic 14
titania 37
sulfides 75, 105
superhydrophilic properties 76, 86
superhydrophobic copper
　foams 34
superhydrophobic films,
　transparent 43
superhydrophobic flower 71
superhydrophobic polymers,
　nanostructured 44
superhydrophobic properties 1–2,
　10, 14–17, 19–20, 24, 28,
　31, 33, 36, 40–42, 49–50,
　52–53, 55–57, 59, 61,
　65–67, 70, 72–73, 75,
　78–79, 81–82, 84–85, 87,
　89–91, 93, 95–96,
　100–101, 104–106,
　109
superhydrophobic/superhy-
　drophilic properties,
　switchable 29–30
superhydrophobic surfaces 1, 12,
　15–16, 65, 74, 101, 104
superhydrophobic walls 6
superhydrophobic ZnO
　nanoplates 80
superhydrophobic ZnS
　nanowires 77
superhydrophobicity 8, 67, 99
superoleophobic properties
　11–14, 16, 34, 38–41,
46, 63–64, 73, 82, 85–89,
　91, 102
surface-enhanced fluorescence
　spectroscopy (SEFS) 24
surface-enhanced Raman
　scattering (SERS) 43, 65,
　79–80, 84, 105

tantalum 45
TEOS, see tetraethoxysilane
tetraethoxysilane (TEOS) 62, 82,
　86, 93
tetramethoxysilane 93
TiO_2 nanoparticles 83, 85, 89
titanium 78
titanium substrates 29–30, 32
　anodizing 46–47
　etching of 46
TMOS, see tetramethoxysilane
tungsten triangular
　nanorods 26

UV illumination 46–47, 77–78
UV irradiation 31, 76, 86, 104–106

water adhesion 4–5, 10, 19, 28,
　36, 40, 42, 59, 78
water striders 2–3
WEDM, see wire electrical
　discharge machining
Wenzel equation 10
Wenzel states 10, 12
wettability 5, 18–19, 46, 60
wire electrical discharge
　machining (WEDM) 18
wrinkles 24

xerogels 93

zinc 16, 18, 69–71
zinc acetate 85–86
zinc substrates 16, 66–67
  etching of 67
ZnCl$_2$ 56, 60

ZnO 15–16, 18, 27, 29, 31, 59, 66, 69–70, 76, 79–80, 83, 87, 99
ZnO nanoparticles 87, 93
ZnO nanorod arrays 60, 71–73, 77, 79
ZnO nanorods 16, 69, 72, 74, 76, 105
ZnO nanowires 73, 78–79